T0396510

Fun Math

**Problem Solving
Beyond the Classroom**

Problem Solving in Mathematics and Beyond

Print ISSN: 2591-7234
Online ISSN: 2591-7242

Series Editor: Dr. Alfred S. Posamentier
Distinguished Lecturer
New York City College of Technology - City University of New York

There are countless applications that would be considered problem solving in mathematics and beyond. One could even argue that most of mathematics in one way or another involves solving problems. However, this series is intended to be of interest to the general audience with the sole purpose of demonstrating the power and beauty of mathematics through clever problem-solving experiences.

Each of the books will be aimed at the general audience, which implies that the writing level will be such that it will not be engulfed in technical language — rather the language will be simple everyday language so that the focus can remain on the content and not be distracted by unnecessarily sophiscated language. Again, the primary purpose of this series is to approach the topic of mathematics problem-solving in a most appealing and attractive way in order to win more of the general public to appreciate this most important subject rather than to fear it. At the same time we expect that professionals in the scientific community will also find these books attractive, as they will provide many entertaining surprises for the unsuspecting reader.

Published

For the complete list of volumes in this series, please visit www.worldscientific.com/series/psmb

Problem Solving in
Mathematics and Beyond · Volume **39**

Fun Math

Problem Solving
Beyond the Classroom

Alfred S. Posamentier

The City University of New York, USA

 **World Scientific**

NEW JERSEY · LONDON · SINGAPORE · BEIJING · SHANGHAI · TAIPEI · CHENNAI

Published by

World Scientific Publishing Co. Pte. Ltd.

5 Toh Tuck Link, Singapore 596224

USA office: 27 Warren Street, Suite 401-402, Hackensack, NJ 07601

UK office: 57 Shelton Street, Covent Garden, London WC2H 9HE

Library of Congress Control Number: 2025003773

British Library Cataloguing-in-Publication Data
A catalogue record for this book is available from the British Library.

Problem Solving in Mathematics and Beyond — Vol. 39
FUN MATH
Problem Solving Beyond the Classroom

ISBN 978-981-12-9744-1 (hardcover)
ISBN 978-981-12-9745-8 (ebook for institutions)
ISBN 978-981-12-9746-5 (ebook for individuals)

For any available supplementary material, please visit
https://www.worldscientific.com/worldscibooks/10.1142/13967#t=suppl

Desk Editors: Kannan Krishnan/Rosie Williamson

Typeset by Stallion Press
Email: enquiries@stallionpress.com

This book is dedicated to all who should learn to love mathematics and to appreciate its beauty and power.

To my children and grandchildren:
Lisa, Daniel, David, Max, Samuel, Jack, and Charles

About the Author

Alfred S. Posamentier is currently Distinguished Lecturer at the New York City College of Technology of the City University of New York. Prior to that, he was Executive Director for Internationalization and Funded Programs at Long Island University, New York. This was preceded by five years as Dean of the School of Education and Professor of Mathematics Education at Mercy University, New York. For the prior 40 years he was at The City College of the City University of New York, where he is now Professor Emeritus of Mathematics Education and Dean Emeritus of the School of Education. He is the author and co-author of more than 90 mathematics books for teachers, secondary and elementary school students, as well as the general readership. Dr. Posamentier is also a frequent commentator in newspapers and journals on topics related to education.

After completing his BA degree in mathematics at Hunter College of the City University of New York, he took a position as a teacher of mathematics at Theodore Roosevelt High School (Bronx, New York), where he focused his attention on improving the students' problem-solving skills and, at the same time, enriching their instruction far beyond what the traditional textbooks offered. During his six-year tenure there, he also developed the school's first mathematics teams (both at the junior and senior levels). He is still involved in working with mathematics teachers and supervisors, nationally and internationally, to help them maximize their effectiveness.

Immediately upon joining the faculty of the City College of New York in 1970 (after having received his master's degree there in 1966), he began to develop in-service courses for secondary school mathematics teachers, including such special areas as recreational mathematics and problem-solving in mathematics. As dean of the City College School of Education for 10 years, his scope of interest in educational issues covered the full gamut of educational issues. During his tenure, as dean he took the school from the bottom of the New York State rankings to the top with a perfect NCATE accreditation assessment in 2009. He also raised more than US$12 million from the private sector for educational innovative programs. Posamentier repeated this successful transition at Mercy University, where he enabled it to become the only college to have received both NCATE and TEAC accreditation simultaneously.

In 1973, Dr. Posamentier received his Ph.D. from Fordham University (New York) in mathematics education and has since extended his reputation in mathematics education to Europe. He has been a visiting professor at several European universities in Austria, England, Germany, the Czech Republic, Turkey and Poland. In 1990, he served as Fulbright Professor at the University of Vienna.

In 1989, he was awarded an Honorary Fellow position at the South Bank University (London, England). In recognition of his outstanding teaching, the City College Alumni Association named him Educator of the Year in 1994, and in 2009. New York City had the day, May 1, 1994 named in his honor by the President of the New York City Council. In 1994, he was also awarded the *Das Grosse Ehrenzeichen für Verdienste um die Republik Österreich* (Grand Medal of Honor from the Republic of Austria), and in 1999, upon approval of Parliament, the President of the Republic of Austria awarded him the title of University Professor of Austria. In 2003, he was awarded the title of *Ehrenbürgerschaft* (Honorary Fellow) of the Vienna University of Technology, and in 2004, he was awarded the *Österreichisches Ehrenkreuz für Wissenschaft & Kunst 1.Klasse* (Austrian Cross of

Honor for Arts and Science, First Class) from the President of the Republic of Austria. In 2005, he was inducted into the Hunter College Alumni Hall of Fame, and in 2006, he was awarded the prestigious Townsend Harris Medal by the City College Alumni Association. He was inducted into the New York State Mathematics Educator's Hall of Fame in 2009, and in 2010, he was awarded the coveted Christian-Peter-Beuth Prize from the Technische Fachhochschule – Berlin. In 2017, Posamentier was awarded *Summa Cum Laude nemmine discrepante* by the Fundacion Sebastian, A.C., Mexico City, Mexico.

He has taken on numerous important leadership positions in mathematics education locally. He was a member of the New York State Education Commissioner's Blue Ribbon Panel on the Math-A Regents Exams and the Commissioner's Mathematics Standards Committee, which redefined the Mathematics Standards for New York State, and he also served on the New York City schools' Chancellor's Math Advisory Panel.

Dr. Posamentier is still a leading commentator on educational issues and continues his long-time passion of seeking ways to make mathematics interesting to teachers, students and the general public, – as can be seen from some of his more recent books.

For more information and a list of his publications, see: https://en.wikipedia.org/wiki/Alfred_S._Posamentier.

Contents

Introduction

Most people recalling their school days have negative feelings about their mathematics classes. Many appear proud to have been weak in mathematics, yet they rarely express such resentment towards any other subject matter. Along with language instruction, mathematics is a dominant subject throughout the twelve years of schooling. One must consider the teaching process when reflecting on why math class leaves unpleasant memories. Throughout the world, at all levels of schooling, testing students is a traditional aspect of education, and has led to curriculum reforms, student assessment, and even teacher assessment. Tests dominate all levels of pre-college mathematics education, and concentrating on what will be tested is the top priority for most teachers. This frequent testing limits the degree to which teachers can venture beyond the curriculum to motivate instruction. The field of mathematics offers a variety of intriguing, entertaining, and inspiring aspects that are too often neglected in the school curriculum.

Mathematics tests are objective and clear regarding their content, leaving little to the imagination. Mathematics teachers focus on the curriculum and tests, and students' scores on these tests indicate the extent to which the material was covered and the effectiveness of the teaching. Frequently, mathematics teachers "teach to the test." This methodology leaves little room for showing students the many beauties of mathematics, which could help motivate students, thereby giving them a far more favorable outlook towards the subject matter. Not only are enticing topics omitted, but also the many skills that accompany these extracurricular concepts. This book hopes to inspire in the general readership a greater appreciation for mathematics by offering topics and skills that are too often neglected in the general education program.

The book's first part concerns four basic areas that are integral aspects of the mathematics curriculum, but presents many examples that shed completely new light on these common topics. The problems are presented with detailed solutions — sometimes multiple solutions — that will further guide the reader towards a new appreciation for the subject matter. The first chapter presents a collection of *Arithmetic Novelties*, which demonstrate arithmetic skills that are not only highly motivating but also helpful in solving quantitative challenges. The second chapter, *Logical Novelties*, presents math problems based on logical thinking rather than on specific mathematics concepts. The units there will surely be enlightening as they provide insights into logical thinking that may have been overlooked in math class. The third chapter, *Algebraic Novelties*, is a collection of mathematics problems that demonstrates the power and importance of algebraic skills and relationships. Rather than the rote methods introduced at school, the reader is offered an opportunity to appreciate this important mathematics tool. The fourth chapter, *Geometric Novelties*, demonstrates the beauty inherent to the field of geometry by presenting mind-boggling geometric properties with unexpected relationships. The reader is merely asked to recall the basic tools learned during the high school geometry course and the rest will be a collection of geometric surprises.

The last two sections of the book expose readers to topics in algebra and geometry that have been neglected at the secondary school level largely because of lack of time. Here you will find some truly wonderful and highly motivating aspects of algebra and geometry. For example, in the chapter *Neglected Topics in Algebra*, Diophantine equations will demonstrate how everyday problems can be solved algebraically with a technique absent from the traditional algebra curriculum. The chapter *Neglected Topics in Geometry* introduces geometric theorems proving amazing relationships that are not presented at the secondary school level.

This book intends to convince readers that there is much to know about math outside of the school curriculum, and could be used to sway more folks to love mathematics. Throughout all the mathematics problems presented, the key factor will be problem-solving techniques that will prove useful beyond mathematics in everyday life. In other words, this book is intended to show that you can have fun with mathematics!

Chapter 1

Arithmetic Novelties

The first form of mathematics that most people are exposed to is arithmetic, an important tool for calculating all sorts of quantitative aspects. Arithmetic remains one of the key concepts, but as technology improves, the arithmetic skills that we were taught are being replaced by calculating devices. However, what is missing throughout our education is exposure to the considerable beauty and unusual aspects of arithmetic that allow us to appreciate the subject from another point of view. Many of these novel aspects can facilitate and simplify calculations. Move along through this section and appreciate arithmetic as you perhaps have not had the pleasure of doing so previously.

Arithmetic Novelty 1

Problem: If Charlie makes a profit of 20% on the selling price of a book, what percent profit does he make on the cost of the book?

Solution: Since the profit is $20\% = \frac{1}{5}$ of the selling price, the cost was $\frac{4}{5}$ of the selling price. We may set up the following proportion:

$$\frac{\text{profit}}{\text{cost}} = \frac{\frac{1}{5} \text{ selling price}}{\frac{4}{5} \text{ selling price}};$$

therefore, the

$$\frac{\text{profit}}{\text{cost}} = \frac{1}{4} = 25\%.$$

Arithmetic Novelty 2

Percentage problems have long been the nemesis of students. These problems get particularly unpleasant when multiple percents need to be processed in the same problem. Consider the following problem.

Problem: Wanting to buy a coat, Lisa is faced with a dilemma. Two competing stores next to each other carry the same brand of coat with the same list price, but at two different discount schemes. Store A offers a 10% discount year-round on all its goods, but on this particular day offers an additional 20% on top of their already discounted price. Store B simply offers a discount of 30% on that day in order to stay competitive. How many percentage-points difference is there between the two options open to Lisa?

Solution: At first glance, one might assume there is no difference in price, since 10% + 20% = 30%, yielding the same discount in both cases. A closer assessment would reveal that this is not correct, since in store A only 10% is calculated on the original list price, with the 20% discount calculated on the lower price, while at store B the entire 30% is calculated on the original price. Now, the question to be answered is: what percentage difference is there between the discounts in store A and store B?

The expected procedure is to assume the cost of the coat to be $100, calculate the 10% discount yielding a $90 price, and an additional 20% off the $90 price will bring the price down to $72. In store B, the 30% discount on $100 would bring the price down to $70, giving a discount difference of $2, or in this case, 2% difference. This procedure, although correct and not too difficult, is a bit cumbersome and does not always allow a full insight into the situation.

An interesting algorithm is provided here:

There is a mechanical method for obtaining a single percentage discount (or increase) equivalent to two (or more) successive discounts (or increases).

(1) Change each of the percents involved into decimal form: .20 and .10.

(2) Subtract each of these decimals from 1.00: to get .80 and .90 (for an increase, add to 1.00).

(3) Multiply these differences: $(.80) \cdot (.90) = .72$.

(4) Subtract this number (i.e., .72) from 1.00: to get $1.00 - .72 = .28$ (if the result of step 3 is greater than 1.00, subtract 1.00 from it to obtain the percent of increase).

When we convert .28 back to percent form, we obtain 28%, the equivalent of successive discounts of 20% and 10%. This percentage (28%) differs from 30% by 2%.

This procedure can also be used to combine more than two successive discounts following the same scheme. In addition, successive increases, combined or not combined with a discount, can also be accommodated in this procedure by adding the decimal equivalent of the increase to 1.00 where the discount was subtracted from 1.00 and then continue the procedure in the same way. If the end result is greater than 1.00, then this reflects an overall increase rather than the discount found in the above problem.

This procedure not only streamlines a typically cumbersome situation, but also provides some insight into the overall picture. For example, consider the question "Is it advantageous to the buyer in the above problem to receive a 20% discount and then a 10% discount, or the reverse, 10% discount and then a 20% discount?" The answer to this question is not immediately intuitively obvious. Yet, since the procedure just presented shows that the calculation is merely multiplication, a commutative operation, we soon find that there is no difference between the two.

Arithmetic Novelty 3

Problem: The United States is one of the few countries in the world that still uses the Fahrenheit temperature scale. Therefore, Americans are often required to convert from Celsius into Fahrenheit. The formula $C = \frac{5}{9}(F-32)$ shows the relationship between the centigrade and Fahrenheit temperature scales. When the centigrade (C) scale reads $20°$, what would the Fahrenheit (F) scale read?

Solution: To change this formula to yield Fahrenheit when centigrade is provided, we multiply both sides of the equation by $\frac{9}{5}$, so that we then have $\frac{9}{5}C = F-32$, which enables us to get $F = \frac{9}{5}C+32$. This equation is a conversion formula in the reverse direction, as it converts C units to F units. Thus, if $C = 20°$, then $F = \frac{9}{5} \cdot 20 + 32 = 68°$.

Arithmetic Novelty 4

Problem: When given the ordered set {5, 6, 7, 8, x}, where x is equal to the average of the three centered elements, what is the average of all five elements?

Solution: The three centered elements are 6, 7, and 8. The average of these three is simply the middle element, namely, 7. This can also be found in the following way:

$$x = \frac{6+7+8}{3} = \frac{21}{3} = 7.$$

Then the average of *all* the elements is

$$\frac{5+6+7+8+7}{5} = \frac{33}{5} = \frac{66}{10} = 6.6.$$

Arithmetic Novelty 5

Problem: Of the following fractions, $\frac{72}{5}, \frac{72}{750}, \frac{12}{5}, \frac{12}{15}$, and $\frac{12}{30}$, what is the smallest proper fraction that will divide each of the following fractions exactly: $\frac{6}{5}, \frac{3}{10}$, and $\frac{4}{15}$?

Solution: In order for one number to divide another number exactly, the quotient must be an integer (whole number). The only choices which will be exactly divided by the listed fractions are $\frac{72}{5}$ and $\frac{12}{5}$. The smaller of these is $\frac{12}{5}$, as

$$\frac{12}{5} \div \frac{6}{5} = 2; \frac{12}{5} \div \frac{3}{10} = 8; \frac{12}{5} \div \frac{4}{15} = 9.$$

Arithmetic Novelty 6

Problem: When Barbara walks to school each day she finds that she can walk 352 steps per minute. She has measured the length of her step to be $1\frac{1}{2}$ feet long. What is Barbara's speed in miles per hour?

Solution: Begin by representing the speed in steps per minute: $\frac{352 \text{ steps}}{1 \text{ minute}}$. Changing steps to feet, we get $\frac{(352)(\frac{3}{2}) \text{feet}}{1 \text{ minute}}$. Considering that there are 5280 feet per mile, we can now change this to miles per hour:

$$\frac{\dfrac{(352)\left(\dfrac{3}{2}\right)}{5280} \text{ miles}}{\dfrac{1}{60} \text{ hour}} = \frac{(352)\left(\dfrac{3}{2}\right)(60)}{5280} = 6 \frac{\text{miles}}{\text{hours}}.$$

Arithmetic Novelty 7

Problem: What number times $\sqrt{.04}$ is equal to 1?

Solution: To find the number that multiplied by $\sqrt{.04}$ will yield 1, we simply consider the fraction $\frac{1}{\sqrt{.04}}$, since $\frac{1}{\sqrt{.04}} \cdot \sqrt{.04} = 1$. We need to determine the value of $\frac{1}{\sqrt{.04}} = \frac{1}{.2}$. When we multiply numerator and denominator by 10, we get

$$\frac{1}{.2} \cdot \frac{10}{10} = \frac{10}{2} = 5.$$

Arithmetic Novelty 8

Problem: Daniel sold two books for 6 dollars each. He gained 20% on the first book he sold, while on the second book he sold he lost 20%. How much did he gain or lose on this transaction of book sales?

Solution: By gaining 20% on the sale of the first book, Daniel had the cost of the first book plus 20% of that cost, which is equal to $6. Therefore, the first book cost him $5, since 20% of the $5 is $1, which would give him the $6 sale.

On the second sale he lost 20% of the sale for $6, which implies that the second book must have cost $7.50, since a 20% reduction from $7.50 is $1.50. Therefore, Daniel paid $6 + $7.50 = $12.50 for the two books and sold them for $12, which would indicate a loss of $.50.

Arithmetic Novelty 9

Problem: What is the sum of the following?

$$(211 \times 555) + (445 \times 789) + (555 \times 789) + (211 \times 445)$$

Solution: To facilitate this apparently complicated computation, factor the expression in the following way:

$$[(211 \times 555) + (211 \times 445)] + [(445 \times 789) + (555 \times 789)]$$
$$= 211 (555 + 445) + 789 (555 + 445)$$
$$= 211 (1000) + 789 (1,000)$$
$$= (211 + 789) (1,000)$$
$$= (1000) (1000) = 1,000,000.$$

Arithmetic Novelty 10

Problem: Evelyn saved $2.50 on buying a pair of gloves that were on sale. She actually spent $25 for the pair of gloves. What percent did she save?

Solution: The original price of Evelyn's gloves was $25.00 + $2.50 = $27.50.

The part of the original price that Evelyn saved is $\frac{\$2.50}{\$27.50} = \frac{1}{11} \approx .09$, or about 9%.

Arithmetic Novelty 11

Problem: A 5-lb box of sugar previously cost $4.50. A 2-lb box of sugar now costs $2.50. What is the ratio of the present price per pound to the previous price per pound?

Solution: The old price per pound was $\frac{4.50}{5} = 0.9$. The present price per pound is $\frac{2.50}{2} = 1.25$. The ratio of the present price per pound to the old price per pound is $\frac{1.25}{0.9} = \frac{125}{90} = \frac{25}{18}$, or $25:18$.

Arithmetic Novelty 12

Problem: The number 63 is $\frac{7}{8}$ of which number?

Solution: This could be interpreted as saying that the number 63 is comprised of 7 units size 9, or to put it more mathematically, $\frac{1}{7}$ of 63 is 9. Therefore, the number which is $\frac{8}{7} \times 63 = 8 \times 9 = 72$.

Arithmetic Novelty 13

Problem: If the average of three fractions is $\frac{1}{4}$, and two of the fractions are $\frac{1}{3}$ and $\frac{1}{4}$, what is the third fraction?

Solution: Let x = the "other" fraction. The average of the three fractions is $\frac{1}{4} = \frac{1}{3}\left[\frac{1}{3} + \frac{1}{4} + x\right]$. Multiply both sides of the equation by 12 to get $3 = 4\left[\frac{1}{3} + \frac{1}{4} + x\right] = \frac{4}{3} + 1 + 4x$. Then multiply both sides by 3 to get: $9 = 4 + 3 + 12x$, so that $x = \frac{1}{6}$.

Arithmetic Novelty 14

Problem: What percent of 2 hours is 5 seconds?

Solution: There are 60 seconds in 1 minute and 60 minutes in 1 hour; therefore, there are $60 \cdot 60$ seconds in 1 hour. Thus, we can represent the 5 seconds as part of the 2 hours represented in seconds as $\frac{5\,\text{seconds}}{2\cdot60\cdot60\,\text{seconds}} \cdot 100\% = \frac{5}{72}\%$.

Arithmetic Novelty 15

Problem: Ernie buys 100 40¢ stamps and 100 50¢ stamps. At the same time, Alice buys 500 40¢ stamps and 500 50¢ stamps. What is the difference between the average spent per stamp by Ernie and Alice?

Solution: One could reason that since they bought an equal number of 40¢ and 50¢ stamps the average per stamp should be the same. This can be calculated as the average spent per stamp by $\text{Ernie} = \frac{40(100)+50(100)}{200} = 45¢.$

The average spent per stamp by

$$\text{Alice} = \frac{40(500)+50(100)}{1,000} = 45¢.$$

Hence, the difference between these averages is zero.

Arithmetic Novelty 16

Problem: If we define the arithmetic operation * for positive numbers to be $a*b = \frac{ab}{a+b}$, then what is the value of 16 * (16 * 16)?

Solution: We can start off by considering first the parenthetical expression $(16*16) = \frac{16\cdot16}{16+16} = \frac{16\cdot16}{2\cdot16} = 8.$ We now need to find $16*8 = \frac{16\cdot8}{16+8} = \frac{128}{24} = 5\frac{1}{3}.$

Arithmetic Novelty 17

Problem: At a "dollar sale," you can buy three items at retail price and get a fourth for a dollar. If Carol buys twelve $10 cans of mixed vegetables, how much does she save during the sale?

Solution: Under normal circumstances Carol would pay $120 for the 12 cans, or $40 for 4 cans. However, during the sale, 4 cans would cost her $30 + $1 = $31. Therefore, for 12 cans she would pay $3 \times \$31 = \$93$, which is a saving of $120 - $93 = $27.

Arithmetic Novelty 18

Problem: There are times in arithmetic where we simply have to compare the sizes of two numbers, as is the case in the following: compare the sizes of $8^{\frac{2}{3}}$ and $32^{\frac{2}{5}}$.

Solution 1: Raise each quantity to the 15th power (the lowest common denominator of the two exponents):

$$\left(8^{\frac{2}{3}}\right)^{15} = 8^{10} = \left(2^3\right)^{10} = 2^{30},$$

$$\left(32^{\frac{2}{5}}\right)^{15} = 32^6 = \left(2^5\right)^6 = 2^{30}.$$

Therefore, we find that the two original quantities are equal.

Solution 2: This technique will simply evaluate the value of each of the numbers and we find that once again they are equal.

$$8^{\frac{2}{3}} = \left(8^{\frac{1}{3}}\right)^2 = 2^2 = 4,$$

$$32^{\frac{2}{5}} = \left(32^{\frac{1}{5}}\right)^2 = 2^2 = 4.$$

Again, we see that the two original quantities must be equal.

Solution 3: Since $8 = 2^3$, we have $8^{\frac{2}{3}} = \left(2^3\right)^{\frac{2}{3}} = 2^2 = 4$.

Similarly, $32 = 2^5$; therefore, $32^{\frac{2}{5}} = \left(2^5\right)^{\frac{2}{5}} = 2^2 = 4$.

Arithmetic Novelty 19

Problem: Comparing powers in fractions requires sometimes looking beyond distractions. Compare the values of $\frac{1}{x^3}$ and $\frac{1}{x^2}$ when $x < 0$.

Solution: When a *negative* number is raised to an *odd* power, the result is *negative*; and when a *negative* number is raised to an *even* power, the result is *positive*. Since x is negative $(x < 0)$, x^3 is also negative, and therefore, $\frac{1}{x^3}$ is negative. Since x^2 is positive, $\frac{1}{x^2}$ is also positive. Thus, we can conclude that $\frac{1}{x^3} < \frac{1}{x^2}$.

Arithmetic Novelty 20

Problem: Find the value of

$$\frac{52{,}340}{(12{,}346)^2 - (12{,}345)(12{,}347)}.$$

Solution: Using a calculator is how we would solve this problem today. However, you never know when you may require some cleverness to sort out an arithmetic challenge without a calculator. The first thing to notice is that the denominator has numbers that are very similar and differ by 1 or 2 units. We can use very simple algebra to make this a trivial problem.

　　If we let $n = 12346$, then we can write the denominator in terms of n as $n^2 - (n-1)(n+1) = n^2 - (n^2 - 1) = 1$. With the value of the denominator as 1, the value of the above fraction is simply the numerator, 52,340.

Arithmetic Novelty 21

Problem: If a apples sell for d dollars, what is the price in cents of b apples selling at the same rate?

Solution: This problem is most effectively solved by setting up a proportion. However, as we begin, we must change the number of dollars to cents so that we have $100d$ cents. Letting x be the cost of b apples in cents, we get the following proportion: $\frac{a}{b} = \frac{100d}{x}$, so that $x = \frac{100bd}{a}$.

Arithmetic Novelty 22

Problem: Find the value of the continued fraction:

$$\cfrac{1}{1+\cfrac{1}{2+\cfrac{1}{3+\cfrac{1}{4}}}}.$$

Solution: We will work from the bottom up as follows:

$$\cfrac{1}{1+\cfrac{1}{2+\cfrac{1}{3+\cfrac{1}{4}}}}=\cfrac{1}{1+\cfrac{1}{2+\cfrac{1}{\frac{13}{4}}}}=\cfrac{1}{1+\cfrac{1}{2+\cfrac{4}{13}}}=\cfrac{1}{1+\cfrac{1}{\frac{30}{13}}}=\cfrac{1}{1+\frac{13}{30}}=\cfrac{1}{\frac{43}{30}}=\frac{30}{43}.$$

Arithmetic Novelty 23

Problem: Arrange the fractions $\frac{2}{3},\frac{5}{7},\frac{7}{15}$, and $\frac{8}{11}$ in order from smallest to largest.

Solution: There are various ways of comparing fractions to determine their relative magnitudes. Three methods are described here, and you will have to decide for yourself which seems easiest. But whatever method you use, the first thing to do with a problem of this kind is to see if there is any obvious way of sorting the fractions by size. In this case, you should notice that $\frac{7}{15}$ is less than $\frac{1}{2}$, whereas the other three fractions are greater than $\frac{1}{2}$. This immediately puts $\frac{7}{15}$ first in the desired sequence and eliminates it from further consideration. All we need to do is find the order of the other three.

Method 1: Fractions can be compared by inspection if they have the same denominator. Therefore, one approach is to convert the given fractions to equivalent fractions with a common denominator. We can do this for all three fractions at once, or we can compare them two at a time. In this problem, the three denominators are 3, 7, and 11, and their least common multiple is $3 \cdot 7 \cdot 11$, or 231. It may be better to compare the fractions in pairs than to convert them to this common denominator. Let us begin by comparing $\frac{2}{3}$ and $\frac{5}{7}$:

$$\frac{2}{3} = \frac{14}{21} \qquad \frac{5}{7} = \frac{15}{21}.$$

Clearly, $\frac{5}{7}$ is larger than $\frac{2}{3}$ (larger by $\frac{1}{21}$).

Now compare $\frac{5}{7}$ with $\frac{8}{11}$:

$$\frac{5}{7} = \frac{55}{77} \qquad \frac{8}{11} = \frac{56}{77}.$$

Clearly, $\frac{8}{11}$ is larger than $\frac{5}{7}$.

We were lucky; if $\frac{8}{11}$ had turned out to be smaller than $\frac{5}{7}$, an additional comparison with $\frac{2}{3}$ would have been necessary. As it is, since $\frac{8}{11}$ is larger than $\frac{5}{7}$, and $\frac{5}{7}$ is larger than $\frac{2}{3}$ (and all of them are larger than $\frac{7}{15}$), the size order is: $\frac{7}{15}, \frac{2}{3}, \frac{5}{7}, \frac{8}{11}$.

Method 2: Another method, which may be easier for those who would rather use a calculator, is to convert all the fractions to decimals by dividing the numerator by the denominator. Again, we can ignore $\frac{7}{15}$, as it is clearly smaller than the rest of the fractions.

$$\frac{2}{3} = \cdot 666666666$$

$$\frac{5}{7} = \cdot 714285714$$

$$\frac{8}{11} = \cdot 727272727$$

The correct order is now obvious: $\frac{7}{15}, \frac{2}{3}, \frac{5}{7}, \frac{8}{11}$.

Method 3: Here we can develop an automatic procedure for comparing fractions. If we have two fractions, $\frac{a}{b}$ and $\frac{c}{d}$, we can convert them to a common denominator as follows:

$$\frac{a}{b} \cdot \frac{d}{d} = \frac{ad}{bd} \qquad \frac{c}{d} \cdot \frac{b}{b} = \frac{cb}{db}.$$

From this, we can see that if a, b, c, and d are positive, $\frac{a}{b} > \frac{c}{d}$ if and only if $ad > bc$. This gives us a direct way of comparing two positive fractions by simply cross-multiplying numerators and denominators and comparing the results. Let us apply this method to $\frac{2}{3}$ and $\frac{5}{7}$.

Since $5 \cdot 3$ is greater than $2 \cdot 7$, we see that the second fraction, $\frac{5}{7}$, must be larger than the first, $\frac{2}{3}$. In similar fashion, we compare $\frac{5}{7}$ and $\frac{8}{11}$ by taking the cross products: $5 \cdot 11 = 55$ and $8 \cdot 7 = 56$. Since $8 \cdot 7$ is greater than $5 \cdot 11$, we then know that $\frac{8}{11}$ is larger than $\frac{5}{7}$.

This method is essentially the same as Method 1, except there we have written the common denominator, but simply compared the numerators.

Arithmetic Novelty 24

Problem: A rectangular photograph whose dimensions are $1\frac{7}{8}$ inches by $2\frac{1}{2}$ inches is to be enlarged so that the larger dimension will be 4 inches long. How long will the shorter dimension become?

Solution: If we let $x = $ length of shorter dimension that we seek, and then set up the following proportion, we will see how compound fractions can be useful.

$$\frac{\text{shorter length}}{\text{longer length}} = \frac{1\frac{7}{8}}{2\frac{1}{2}} = \frac{x}{4}.$$

At this point one can simplify the compound fraction

$$\frac{\frac{15}{8}}{\frac{5}{2}} = \frac{8 \cdot \frac{15}{8}}{8 \cdot \frac{5}{2}} = \frac{15}{5 \cdot 4} = \frac{15}{20} = \frac{3}{4},$$

and that is equal to $\frac{x}{4}$, so that $x = 3$.

Arithmetic Novelty 25

Problem: In a small town, the number of students in elementary schools is expected to be $\frac{1}{5}$ of the population. There will be $\frac{1}{4}$ as many students in high schools as in elementary schools, and $\frac{2}{5}$ as many in colleges as in high schools. What percent of the population is expected to be in colleges?

Solution: In a problem involving a chain of relationships, we may begin at either end of the chain. We let $E =$ the number of elementary students, $H =$ the number of high school students, $C =$ the number college students, and $P =$ the entire population. Therefore, we have the following:

$$E = \frac{1}{5}P, \quad H = \frac{1}{4}E, \quad C = \frac{2}{5}H.$$

We then have:

$$C = \frac{2}{5}H = \frac{2}{5}\left(\frac{1}{4}E\right) = \frac{1}{10}\left(\frac{1}{5}P\right) = \frac{1}{50}P.$$

Since we have $C = \frac{1}{50}P$, this indicates that the college students are $\frac{1}{50}$ of the population, or .02P, or 2% of the population.

Arithmetic Novelty 26

Problem: If it is now 10:45 am, what time will it be 143,999,999,995 minutes later?

Solution: This problem appears to be rather complex because of the 12-digit number 143,999,999,995. Let's examine this problem from another point of view. The number 143,999,999,995 is only five minutes short of being 144,000,000,000 minutes, which is an exact multiple of 60. That is, 144,000,000,000 minutes ÷ 60 = 2,400,000,000 hours = 100,000,000 days. In this case, the time would also be 10:45 am. Since the given number was five minutes short of that, the correct time would be 10:40 am. Here is an example of the obvious advantage of examining the problem from a different point of view.

Arithmetic Novelty 27

Problem: What is the smallest prime number that divides the sum $5^7 + 7^{11} + 11^{13} + 13^5$?

Solution: At first glance, one may decide to attempt to find the actual values for each of the terms: 5^7, 7^{11}, 11^{13}, and 13^5. However, finding the sum of these four terms and testing the various prime divisors will take a great deal of time. There must be a better way to approach this problem. There are no factors of 2 included in this given sum, since we find that 5^7 is odd, 7^{11} is odd, 11^{13} is odd, and 13^5 is odd. The sum of four odd numbers must be even. The smallest prime number that divides the sum will be 2. The problem is easily resolved by examining it from an alternate point of view.

Arithmetic Novelty 28

Problem: The basketball squad is taking part in a free-throw contest. The first player scored x free-throws. The second player scored y free-throws. The third shooter made the same number of free-throws as the arithmetic mean of the number of free-throws made by the first two players. Each subsequent player in the contest scored the arithmetic mean of the number of free-throws made by all the players who had preceded her. How many free-throws did the twentieth player make?

Solution: One may decide to solve this problem by finding the arithmetic mean for each player in turn, using the results of all the previous shooters. This would take much too long!

Let's examine a simplified version of the problem by replacing x and y with simple numbers. Suppose the first player made 8 free throws and the second player made 12. Then the third shooter made a number of free-throws equal to their arithmetic mean, or $\frac{8+12}{2} = 10$. The number of free-throws made by the fourth player is the arithmetic mean of the scores made by the first three players, or $\frac{8+12+10}{3} = 10$. Similarly, the score made by the fifth player is the arithmetic mean of the scores made by the first four players, which is $\frac{8+12+10+10}{4} = 10$. This reveals that the score made by any player after the first two players will be the same as the arithmetic mean for the first two players' scores.

We now return to the original problem. The correct answer is simply the arithmetic mean of the scores made by the first two players, or $\frac{x+y}{2}$. Here we see how a simpler, analogous problem enables us to determine the method for solving the original problem rather quickly.

Arithmetic Novelty 29

Problem: Given 19 consecutive integers whose sum is 95, what is the tenth of these numbers?

Solution: We may approach the problem in a simple manner. Of 19 consecutive integers, the tenth integer is the middle integer, and it is the average (or arithmetic mean) of the given 19 integers. Thus, we simply take the sum, 95, and divide it by the numbers of integers, 19. Thus, the tenth integer is 5.

Arithmetic Novelty 30

Problem: What is the quotient of $\frac{1}{500,000,000,000}$?

Solution: This calculation would even be a challenge for a computer, so we need to approach this calculation from another vantage point. Finding a pattern could lead to a solution, so let's consider the following:

	Number of zeros after 5	The quotient	Number of zeros between the decimal point and the number 2
$\frac{1}{5}$	0	.2	0
$\frac{1}{50}$	1	.02	1
$\frac{1}{500}$	2	.002	2
$\frac{1}{5,000}$	3	.0002	3
$\vdots$	$\vdots$	$\vdots$	$\vdots$
$\frac{1}{500,000,000,000}$	11	.000000000002	11

From the pattern observed in the above table, the quotient is easily calculated.

Arithmetic Novelty 31

Problem: Find the largest three-digit numbers written in base 10 that are divisible by 22 and where the sum of the tens digit and units digit is 11.

Solution: We know immediately that if the number is to be divisible by 22, then it must be divisible by two, that is, it must be an even number. Therefore, the units digit must be either 2, 4, 6, or 8, and the last two digits of this three-digit number would be 92, 74, and 38 so that the sum of their digits is 11. In order for a three-digit number to be divisible by 11, the sum of the hundreds digit and units digit must equal the tens digit. Thus, the only three-digit numbers to qualify are 792, 374, and 638.

Arithmetic Novelty 32

Problem: How many four-digit integers (in base 10 notation) have the property that the sum of the fifth powers of each of their digits is less than 32?

Solution: We need to recognize immediately that the first digit (in the thousandth place) must be 1. Furthermore, the remaining three digits must be either 1 or 0, since a 2 cannot be one of the digits as $2^5 = 32$, which would take the sum beyond the required limit. Therefore, there are two possibilities for each of the remaining three places. Thus, there are $2^3 = 8$ possibilities.

Arithmetic Novelty 33

Problem: Find the numerical value of the following:

$$1 + 2 - 3 - 4 + 5 + 6 - 7 - 8 + 9 + 10 - 11 - 12 + \cdots + 97 + 98 - 99 - 100.$$

Solution: The trick to getting this sum is to group these in fours as follows:

$$(1 + 2 - 3 - 4) + (5 + 6 - 7 - 8) + (9 + 10 - 11 - 12) + \cdots + (97 + 98 - 99 - 100).$$

We then notice that each group of four has a sum of -4. Therefore, since there are 25 such groups, the sum is $(25) \cdot (-4) = -100$.

Arithmetic Novelty 34

Problem: In base 10, the number 80,999? is a prime number, where ? represents the units digit. What is the value of ? in order for the number to be a prime number?

Solution: The units digit cannot be an even number or 5 if this is to be a prime number. Therefore, the only possible units digit that could be considered are 1, 3, 7, or 9. We can eliminate the numbers 1 and 7 because in each case the sum of the digits would be a multiple of 3 and, therefore, the number itself would be a multiple of 3 and not a prime number. If ? = 9, then the number would be $809,999 = 810,000 - 1 = 900^2 - 1 = (900 + 1)(900 - 1) = 901 \cdot 899$, so 809,999 is not a prime number. This leaves us with the only other number remaining amongst the original odd numbers, ? = 3, which results in the prime number 809,993.

Arithmetic Novelty 35

Problem: Given the sequence of integers: 1, 2, 2, 3, 3, 3, 4, 4, 4, 4, 5, 5, 5, 5, 5, ... , where each positive integer, n, occurs in a grouping n consecutive times. How many terms of this sequence are needed so that the sum of the reciprocals is 500?

Solution: At first glance, the problem as posed appears to require some extensive number-crunching. Let's organize data in a slightly different fashion and follow the wording in the original problem; perhaps this will enable us to solve the problem more readily. Consider the sum of the reciprocals:

$$\frac{1}{1}, \frac{1}{2}, \frac{1}{2}, \frac{1}{3}, \frac{1}{3}, \frac{1}{3}, \frac{1}{4}, \frac{1}{4}, \frac{1}{4}, \frac{1}{4}, \frac{1}{5}, \frac{1}{5}, \frac{1}{5}, \frac{1}{5}, \frac{1}{5}, \ldots$$

Notice the pattern that has emerged, as we consider these reciprocals in "clusters":

$$\frac{1}{1} = 1\,(1\text{ term})$$

$$\frac{1}{2} + \frac{1}{2} = 1\,(2\text{ terms})$$

$$\frac{1}{3} + \frac{1}{3} + \frac{1}{3} = 1\,(3\text{ terms})$$

$$\frac{1}{4} + \frac{1}{4} + \frac{1}{4} + \frac{1}{4} = 1\,(4\text{ terms})$$

$$\frac{1}{5} + \frac{1}{5} + \frac{1}{5} + \frac{1}{5} + \frac{1}{5} = 1\,(5\text{ terms})$$

and so on.

Thus, for every grouping of n consecutive terms in the original series, the sum of the reciprocals equals 1. Now the problem is easily resolved. We are looking for the sum of the first 500 integers from 1 through 500. This is obtained by $\left(\frac{500+1}{2}\right)(500) = 125{,}250$. The problem was easily solved once we had organized the data in a more meaningful manner.

Arithmetic Novelty 36

Problem: Determine the *exact time* that the hands of a clock will overlap after 4:00 o'clock.

Solution: Your first reaction to this problem is likely to be that the answer is simply 4:20. Wrong! That does not take into account that the hour hand moves uniformly while the minute hand moves faster. With this in mind, the astute reader will begin to estimate the answer to be between 4:21 and 4:22. Realize that the hour hand moves through an interval between minute markers every 12 minutes. Therefore, it will leave the interval 4:21 – 4:22 at 4:24. This, however, doesn't answer the original question about the exact time of this overlap.

We will provide a "trick" to deal better with this situation. Realizing that this first guess of 4:20 is not the correct answer, since the hour hand does not remain stationary and moves when the minute hand moves, the trick is to simply multiply the 20 (the wrong answer) by $\frac{12}{11}$ to get $21\frac{9}{11}$, which yields the correct answer: $4:21\frac{9}{11}$.

One way to understand the movement of the hands of a clock is by considering the hands traveling independently around the clock at uniform speeds. The minute markings on the clock (from now on referred to as "markers") will serve to denote distance as well as time. Now consider 4 o'clock as the initial time on the clock. Our problem will be to determine exactly when the minute hand will overtake the hour hand after 4 o'clock. Consider the speed of the hour hand and the number of markers traveled that the minute hand must travel to overtake the hour hand. Let us refer to this distance as d markers. Hence the distance that the hour hand travels is $d - 20$ markers, since it has a 20-marker head start over the minute hand. (See Figure 1-1.)

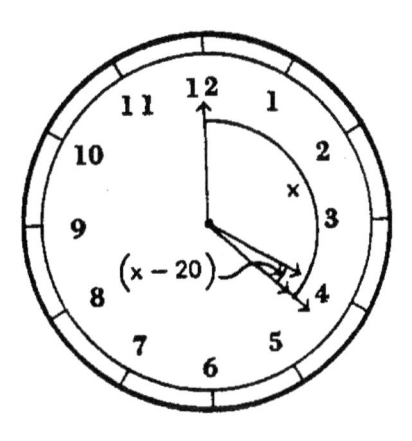

Figure 1-1

For this to take place, the times required for the minute hand, $\frac{d}{12r}$, and for the hour hand, $\frac{d-20}{r}$, are the same. Therefore, $\frac{d}{12r} = \frac{d-20}{r}$, and $d = \frac{12}{11}\cdot 20 = 21\frac{9}{11}$. Thus, the minute hand will overtake the hour hand at exactly $4:21\frac{9}{11}$. Consider the expression $d = \frac{12}{11} \cdot 20$. The quantity 20 is the number of markers that the minute hand had to travel to get to the desired position, if we assume the hour hand remained stationary. However, the hour hand does not remain stationary. Hence, we must multiply this quantity by $\frac{12}{11}$, since the minute hand must travel $\frac{12}{11}$ as far. Let us refer to this fraction, $\frac{12}{11}$, as the correction factor.

To begin to familiarize yourself with the use of the correction factor, choose some short and simple problems. For example, you may seek to find the exact time when the hands of a clock overlap between 7 and 8 o'clock. Here you would first determine how far the minute hand would have to travel from the "12" position to the position of the hour hand, assuming again that the hour hand remains stationary. Then by multiplying the number of markers, 35, by the *correction factor*, $\frac{12}{11}$, you will obtain the exact time, $7{:}38\frac{2}{11}$, that the hands will overlap.

To enhance your understanding of this new procedure, consider a person checking a wristwatch against an electric clock and noticing that the hands on the wristwatch overlap every 65 minutes (as measured by the electric clock). Is the wristwatch fast, slow, or accurate? You may wish to consider the problem in the following way. At 12 o'clock the hands of a clock overlap exactly. Using the previously described method we find that the hands will again overlap at exactly $1{:}05\frac{5}{11}$, and then again at exactly $2{:}10\frac{10}{11}$, and again at exactly $3{:}16\frac{4}{11}$ and so on. Each time there is an interval of $65\frac{5}{11}$ minutes between overlapping positions. Hence, the person's watch is inaccurate by $\frac{5}{11}$ of a minute. Can you now determine if the wristwatch is fast or slow?

There are many other interesting, and sometimes rather difficult, problems made simple by this "correction factor." You may very easily pose your own problems. For example, you may wish to find the exact times when the hands of a clock will be perpendicular (or form a straight angle) between, say, 8 and 9 o'clock. Again, you would try to determine the number of markers that the minute hand would have to travel from the "12" position until it forms the desired angle with the stationary hour hand. Then multiply this number by the correction factor, $\frac{12}{11}$, to obtain the exact actual time. That is, to find the exact time that the hands of a clock are *first* perpendicular between 8 and

9 o'clock, determine the desired position of the minute hand when the hour hand remains stationary (here, on the 25-minute marker). Then, multiply 25 by $\frac{12}{11}$ to get 8:27$\frac{3}{11}$, the exact time when the hands are *first* perpendicular after 8 o'clock.

For those who want to look at this issue from a non-algebraic viewpoint, you could justify the $\frac{12}{11}$ correction factor for the interval between overlaps in the following way. Think of the hands of a clock at noon. During the next 12 hours (i.e., until the hands reach the same position at midnight) the hour hand makes one revolution, the minute hand makes 12 revolutions, and the minute hand coincides with the hour hand 11 times (including midnight, but not noon, starting just after the hands separate at noon). Since each hand rotates at a uniform rate, the hands overlap each $\frac{12}{11}$ of an hour, or 65$\frac{5}{11}$ minutes. This can be extended to other situations. You should derive a great sense of achievement and enjoyment from employing this simple procedure to solve what appears to be a very difficult clock problem.

Arithmetic Novelty 37

Problem: If a phonograph record can make 33$\frac{1}{3}$ revolutions per minute, in how many seconds will the record revolve 180°?

Solution: Let x represent the number of seconds that are required for the record to revolve 180°, or one-half a revolution. We then set up the following proportion:

$$\frac{x \text{ seconds}}{60 \text{ seconds}} = \frac{\frac{1}{2} \text{ revolutions}}{33\frac{1}{3} \text{ revolutions}}.$$

From this we derive the following equation: $\frac{1}{2} \cdot 60 = 33\frac{1}{3}x$, which becomes $\frac{100x}{3} = 30$. So the record makes a one-half revolution in $\frac{9}{10}$ seconds.

Arithmetic Novelty 38

Problem: In numbering the pages of a book, 3,357 digits are required. How many pages are in this book?

Solution: To determining the number pages in the book, we will consider them in categories of size as follows:

- Pages 1–9 use 9 digits.
- Pages 10–99 use (90)(2) digits, or 180 digits.
- Pages 100–999 use (900)(3) digits, or 2,700 digits.

At this point, $2,700 + 180 + 9 = 2889$ digits of the 3,357 digits have been used to number pages 1–999. The remaining are 468 digits, which can be used to form 117 four-digit numbers. Therefore, the total number of pages is $999 + 117 = 1,116$.

Arithmetic Novelty 39

Problem: If $\frac{5}{8}$ of a gallon of liquid soap costs \$2.75, how much will $\frac{3}{5}$ of a gallon of that same liquid soap cost?

Solution: Set up the following proportion:

$$\frac{\frac{5}{8} \text{ gallon}}{\frac{3}{5} \text{ gallon}} = \frac{\$2.75}{x} \left(\text{where } x = \text{cost of } \frac{3}{5} \text{ gallon of soap} \right)$$

$$\frac{\frac{5}{8} \text{ gallon}}{\frac{3}{5} \text{ gallon}} \cdot \frac{40}{40} = \frac{25}{24} = \frac{\$2.75}{x}.$$

Cross multiply the two above fractions so that $25x = 24 \cdot \$2.75$, and we get $x = \$2.64$.

Arithmetic Novelty 40

Problem: In a certain state, a car is taxed according to its horsepower as follows: 3 dollars for each of the first 24 horsepower and 4 dollars for each succeeding horsepower. What is the horsepower of a car whose tax is $100?

Solution: The tax on the first 24 horsepower = ($3)(24) = $72. The remaining money ($28) is absorbed by 7 additional horsepower at the rate of $4 each.

Therefore, the total horsepower = 24 + 7 = 31.

Arithmetic Novelty 41

Problem: Find the next three terms in the following sequence: 0, 1, 2, 3, 6, 7, 14, 15, 30, ___, ___, ___.

Solution: The sequence follows a pattern, where the three sought after numbers are bold:

Computation of sequence	2(0)	2(0)+1	2(1)	2(1)+1	2(3)	2(3)+1	2(7)	2(7)+1	2(15)	**2(15)+1**	**2(31)**	**2(31)+1**
Sequence	0	1	2	3	6	7	14	15	30	**31**	**62**	**63**

Arithmetic Novelty 42

Problem: A high-speed train travels 1000 miles in 5 hours. How many miles per hour would this train have to increase its rate to complete the trip in $4\frac{1}{2}$ hours?

Solution: If it takes a train 5 hours to travel 1,000 miles, its rate is $\frac{1000}{5}$, or 200 miles per hour. Similarly, if it takes $4\frac{1}{2}$ hours to travel 1000 miles, its rate is $\frac{1000}{4\frac{1}{2}}$, or $222\frac{2}{9}$ miles per hour, which is an increase of $22\frac{2}{9}$ miles per hour.

Arithmetic Novelty 43

Problem: The cost of a calculator is 18 dollars, and the selling price is S dollars. What is the percent profit on the selling price?

Solution: The profit earned by selling a calculator is $(S - 18)$ dollars. The *part* of the selling price which is profit is $\frac{S-18}{S}$. Then the percent of the selling price which is profit is $\frac{100(S-18)}{S}$.

Arithmetic Novelty 44

Problem: Two competitive stores change the discount rates on a regular basis. On one particular day, store A offers a 10% discount on all items. However, when store B sees this offer, it offers a 15% discount on all of its items, whereupon store A decides to top the competitor by offering an additional 20% on all of its items. To remain competitive and not take a loss, store B decides to offer an additional 15% on all its items. It would appear that the discounts at both stores are now 30%. However, this is not the case. Which store offers a better discount?

Solution: The technique for solving this problem was presented as Arithmetic Novelty 2, and it is offered here to see if the technique can be recalled. In order to find one discount equivalent to two successive discounts, subtract each of the discounts (in decimal form) from 1.00, multiply the resulting differences, and subtract this product from 1.00. For example, the one discount equivalent to successive discounts of 10% and 20% is found in the following way:

$$\left.\begin{array}{l} 1.00 - .10 = .90 \\ 1.00 - .20 = .80 \end{array}\right\} \text{multiply to get} : .72$$

then : $1.00 - .72 = .28$ or $\underline{\underline{28\%}}$.

In a similar way, we can find one discount equivalent to two successive discounts of 15% and 15% by subtracting each from 100%, multiplying results and subtracting that from 100%, that is, $(100 - .15)(100 - .15) = (.85)(.85) = .7225$, and then

$100 - .7225 = .2775 = 27.75\%$, the combined discount for store B. Therefore, the discount at A exceeds the discount at B by $.25\%$ or $\frac{1}{4}\%$.

Arithmetic Novelty 45

Problem: A dinner check was to be divided equally among 8 people. When one person refused to pay, each of the others had to pay $20 more per person. What was the total amount of the check?

Solution: Let x represent the total number of dollars of the check. Then $\frac{x}{8}$ represents the number of dollars each person was to have originally paid, while $\frac{x}{7}$ represents the number of dollars each person paid in the second case. Therefore, $\frac{x}{8} + 20 = \frac{x}{7}$, so that $x = \$1,120$, which is the amount of the original check.

Arithmetic Novelty 46

Problem: In a theater with 396 people, the ratio of women to men was 2:3, and the ratio of men to children was 1:2. How many men were in the theater?

Solution: Since the ratio of women to men is 2:3 and the ratio of men to children is 1:2 (or 3:6), the ratio of women to men to children is 2:3:6. Hence: $2x + 3x + 6x = 396$, and $x = 36$. Therefore, the number of men $= 3x = 108$.

Arithmetic Novelty 47

Problem: Find the simple value of the following continued fraction:

$$\cfrac{1}{2 - \cfrac{1}{2 - \cfrac{1}{2 - \cfrac{1}{2 - \cfrac{1}{2}}}}}.$$

Solution: Follow the step-by-step process:

$$\cfrac{1}{2-\cfrac{1}{2-\cfrac{1}{2-\cfrac{1}{2}}}} = \cfrac{1}{2-\cfrac{1}{2-\cfrac{1}{\frac{3}{2}}}} = \cfrac{1}{2-\cfrac{1}{2-\frac{2}{3}}} = \cfrac{1}{2-\frac{1}{\frac{4}{3}}} = \cfrac{1}{2-\frac{3}{4}} = \cfrac{1}{\frac{5}{4}} = \frac{4}{5}.$$

Arithmetic Novelty 48

Problem: Which of these two fractions is larger? $\frac{x}{y}$, or $\frac{x+k}{y+k}$, when $x > y > 0$, and $k > 0$.

Solution: The easiest way to solve this problem is to compare the cross products $x(y + k)$ and $y(x + k)$, which equal $xy + xk$ and $xy + yk$, respectively. Since k is positive and $x > y$, we have $xk > yk$, which indicates that the first cross-product is the greater. Therefore, $\frac{x}{y} > \frac{x+k}{y+k}$.

Arithmetic Novelty 49

Problem: In a certain city the population grew by 3% from 2023 to 2024. From 2024 to 2025, the population increased by 6%. What was the population growth for the period from 2023 to 2025?

Solution: Let P represent the population in 2023. Then $P + .03P$ is the population in 2024. Then the population in 2025 can be represented as

$$(P + .03P) + .06(P + .03P) = (P + .03P)\,(1.06)$$
$$= (1.03)\,P(1.06) = 1.0918P,$$

which represents an increase in population of 9.18%.

An alternate solution would be to see the problem as an application of successive discounts (see Arithmetic Novelties 2 and 45), where you merely multiply $(1.03) \cdot (1.06) = 1.0918$, which indicates an increase of 9.18%.

Arithmetic Novelty 50

Problem: Find the value of $\sqrt[4]{4^{11}+4^{11}+4^{11}+4^{11}}$.

Solution: This can be rewritten as $(4\cdot4^{11})^{\frac{1}{4}}=(4^{12})^{\frac{1}{4}}=4^3=64$.

Arithmetic Novelty 51

Problem: If $\frac{24}{19}=a+\cfrac{1}{b+\cfrac{1}{c+\frac{1}{d}}}$, find the value of *d.*

Solution: The process of breaking this down into continued fractions goes as follows:

$$\frac{24}{19}=1+\frac{5}{19}=1+\cfrac{1}{\frac{19}{5}}=1+\cfrac{1}{3+\cfrac{4}{5}}=1\cfrac{1}{3+\cfrac{1}{\frac{5}{4}}}=1+\cfrac{1}{3+\cfrac{1}{1+\cfrac{1}{4}}}.$$

Therefore, $d=4$.

Arithmetic Novelty 52

Problem: The rational fraction in lowest terms $\frac{N}{D}$ is equivalent to the infinitely repeating decimal $1.266666\overline{6}$. Find the fraction $\frac{N}{D}$.

Solution: Rather than puzzle out the fraction which yields the given infinitely repeating decimal, we shall work backwards by "dismantling" the decimal expression, which is the traditional way of converting infinitely repeating decimals to common fractions. In this case, let $x=.06666\overline{6}$. Then $10x=.06666\overline{6}$, and $100x=6.6666\overline{6}$. By subtraction we get $100x-10x=90x=6$, and $x=\frac{6}{90}$.

We are now ready to consider the original problem.

$$1.266666\overline{6}=1.2+0.066666\overline{6}\cdot=\frac{6}{5}+\frac{6}{90}=\frac{(6)(18)}{(5)(18)}+\frac{6}{90}=\frac{(6)(19)}{90}=\frac{19}{15}.$$

Arithmetic Novelty 53

Problem: Find the value of the following (without a calculator, of course!).

$$\frac{10^2 + 11^2 + 12^2 + 13^2 + 14^2}{365}.$$

Solution: Rather than to do the indicated arithmetic, consider adopting a different point of view, that is, rewrite each of the terms as follows:

$$\frac{(12-2)^2 + (12-1)^2 + 12^2 + (12+1)^2 + (12+2)^2}{365}.$$

By expanding each of the binomial squares we get:

$$\frac{(12^2 - 48 + 4) + (12^2 - 24 + 1) + 12^2 + (12^2 + 24 + 1) + (12^2 + 48 + 4)}{365}.$$

This can be simplified by combining terms as follows:

$$\frac{5(12^2) + 4 + 1 + 1 + 4}{(5)(73)}.$$

This gives us:

$$\frac{720 + 10}{(5)(73)} = 2.$$

Although a calculator computation would have been just as efficient, this is perhaps more elegant!

Arithmetic Novelty 54

Problem: Find the numerical value of the following expression:

$$\left(1-\frac{1}{4}\right)\left(1-\frac{1}{9}\right)\left(1-\frac{1}{16}\right)\left(1-\frac{1}{25}\right)\cdots\left(1-\frac{1}{225}\right).$$

Solution: When faced with this problem it is wise to simplify each of the 14 parentheses' expressions to get:

$$\left(\frac{3}{4}\right)\left(\frac{8}{9}\right)\left(\frac{15}{16}\right)\left(\frac{24}{25}\right)\cdots\left(\frac{224}{225}\right).$$

When you try to change each fraction to a decimal (with a calculator) and multiply the results (again with a calculator), it is obviously a very cumbersome calculation.

An alternative method would be to organize the data in a different way. This will permit us to look at the problem from a different point of view and see some sort of pattern that will enable us to simplify our work.

$$\left(1^2-\frac{1}{2^2}\right)\left(1^2-\frac{1}{3^2}\right)\left(1^2-\frac{1}{4^2}\right)\left(1^2-\frac{1}{5^2}\right)\cdots\left(1^2-\frac{1}{15^2}\right).$$

We now factor each parenthesis as the difference of two perfect squares, which yields:

$$\left(1-\frac{1}{2}\right)\left(1+\frac{1}{2}\right)\left(1-\frac{1}{3}\right)\left(1+\frac{1}{3}\right)\left(1-\frac{1}{4}\right)\left(1+\frac{1}{4}\right)\left(1-\frac{1}{5}\right)\left(1+\frac{1}{5}\right)\cdots$$

$$\left(1-\frac{1}{14}\right)\left(1+\frac{1}{14}\right)\left(1-\frac{1}{15}\right)\left(1+\frac{1}{15}\right)$$

$$=\left(\frac{1}{2}\right)\left(\frac{3}{2}\right)\left(\frac{2}{3}\right)\left(\frac{4}{3}\right)\left(\frac{3}{4}\right)\left(\frac{5}{4}\right)\left(\frac{4}{5}\right)\left(\frac{6}{5}\right)\cdots\left(\frac{13}{14}\right)\left(\frac{15}{14}\right)\left(\frac{14}{15}\right)\left(\frac{16}{15}\right).$$

A pattern is now evident, and we may "cancel" throughout the expression. And as a result, we are left with $\left(\frac{1}{2}\right)\left(\frac{16}{15}\right)=\frac{8}{15}$.

Arithmetic Novelty 55

Problem: We are given the following four numbers

$$7,895$$
$$13,127$$
$$51,873$$
$$7,356.$$

What percent of their sum is their average?

Solution: The typical respondent does exactly what the problem calls for. That is, first find the sum of the four numbers, then find their average, and finally, divide and convert to a percent.

We can solve a simpler analogous problem by considering the general case. Let the sum of these numbers be represented by S. Their average is $\frac{S}{4}$. Now, to find what percent the average is of the sum, we first divide $\frac{\frac{S}{4}}{S} = \frac{1}{4}$. We now change $\frac{1}{4}$ to a percent, which is 25%.

Arithmetic Novelty 56

Problem: For consecutive, positive, odd integers a, b, and c (where $a < b < c$), is the sign of the expression $(c-a)(b-a)(c-b)(a-c)(a-b)(b-c)$ positive or negative?

Solution: You may wish to actually multiply the six factors and obtain the product. One would substitute values for a, b, and c (say, 15, 17, 19) to obtain the sign of the expression.

However, we can solve this problem by solving a simpler analogous problem. We start by selecting some conveniently small, consecutive, positive, odd integers to replace a, b, and c. Let's choose the simplest: $a = 1, b = 3, c = 5$.

Then, $(c-a)(b-a)(c-b)(a-c)(a-b)(b-c)$ would be represented by

$$(5-1)(3-1)(5-3)(1-5)(1-3)(3-5) =$$

$$(4) \quad (2) \quad (2) \quad (-4) \quad (-2) \quad (-2) = -256,$$

Thus, the sign of the expression is negative.

An alternative solution is to note that $(a - c)$ is the negative of $(c - a)$, $(a - b)$ and $(b - a)$ are negatives of one another, and so are $(b - c)$ and $(c - b)$. Therefore, there are three negative factors in the product, ensuring the product to be negative.

Note that this problem could also have been solved in the same way if the given numbers were not necessarily consecutive, just all three different.

Arithmetic Novelty 57

Problem: The divisors of 360 add up to 1,170. What is the sum of the reciprocals of the divisors of 360?

Solution: The most obvious solution would be to find all the divisors of 360, take their reciprocals, and then add them. The divisors of 360 are 1, 2, 3, 4, 5, 6, 8, 9, ... , 120, 180, 360. The reciprocals are $\frac{1}{1}, \frac{1}{2}, \frac{1}{3}, \frac{1}{4}, \frac{1}{5}, \frac{1}{6}, \frac{1}{8}, \frac{1}{9}, \ldots, \frac{1}{120}, \frac{1}{180}, \frac{1}{360}$. We now find the common denominator (360), convert all the fractions to their equivalents, and add. It is quite easy to make a mechanical or computational error, or to miss one or more divisors.

To examine a simpler analogous problem, find the sum of the reciprocals of the divisors of 12 and see if this helps. The divisors of 12 are 1, 2, 3, 4, 6, and 12. Their sum is $1 + 2 + 3 + 4 + 6 + 12 = 28$. Now let's find the sum of the reciprocals of these factors.

$$\frac{1}{1} + \frac{1}{2} + \frac{1}{3} + \frac{1}{4} + \frac{1}{6} + \frac{1}{12} = \frac{28}{12}.$$

Aha! The numerator of the fraction is the sum of the divisors, while the denominator is the number we're working with.

Now we can solve our original problem. The sum of the factors of 360 is 1,170. Thus, the sum of the reciprocals of the factors must be $\frac{1,170}{360}$.

Arithmetic Novelty 58

Problem: Find the value of

$$\frac{2+4+6+8+\cdots+34+36+38}{3+6+9+12+\cdots+51+54+57}.$$

Solution: The traditional approach, and one that many people would use, is to add the nineteen numbers in the numerator and the nineteen numbers in the denominator using a calculator. They would then divide the two sums to find the value of the fraction (again using their calculators). This solution is a valid one and should yield the correct answer. However, it requires a great deal of effort, and could lead to a mistake.

Instead, consider solving a simpler problem and see if it leads us to the answer to the original problem. We'll begin with one term in the numerator and denominator, then expand to two terms in each, then three, and so on.

$$\frac{2}{3}=\frac{2}{3},\quad \frac{2+4}{3+6}=\frac{6}{9}=\frac{2}{3},\quad \frac{2+4+6}{3+6+9}=\frac{12}{18}=\frac{2}{3},\quad \frac{2+4+6+8}{3+6+9+12}=\frac{20}{30}=\frac{2}{3}.$$

The answer is always $\frac{2}{3}$.

An alternate solution also requires solving a simpler problem by getting the problem in a simpler form. This would be a factored form:

$$\frac{2(1+2+3+\cdots+17+18+19)}{3(1+2+3+\cdots+17+18+19)}=\frac{2}{3}.$$

Arithmetic Novelty 59

Problem: Given the following numbers $2^{\frac{1}{2}},3^{\frac{1}{3}},8^{\frac{1}{8}},9^{\frac{1}{9}}$, which of these four numbers is the largest and which of these four numbers is the second largest?

Solution: Since $\left(2^{\frac{1}{2}}\right)^8 = 2^4 = 16$, and $\left(8^{\frac{1}{8}}\right)^8 = 8$, we have $2^{\frac{1}{2}} > 8^{\frac{1}{8}}$. Also, since $\left(2^{\frac{1}{2}}\right)^{18} = 2^9 = 512$, and $\left(9^{\frac{1}{9}}\right)^{18} = 9^2 = 81$, we then have $2^{\frac{1}{2}} > 8^{\frac{1}{8}}$. Furthermore, since $\left(3^{\frac{1}{3}}\right)^6 = 3^2 = 9$, and $\left(2^{\frac{1}{2}}\right)^6 = 8$, we have $3^{\frac{1}{3}} > 2^{\frac{1}{2}}$.

Since $3^{\frac{1}{3}} > 2^{\frac{1}{2}}$ and $2^{\frac{1}{2}} > 8^{\frac{1}{8}}$ and $2^{\frac{1}{2}} > 9^{\frac{1}{9}}$, the largest of these four numbers is $3^{\frac{1}{3}}$ and the next largest is $2^{\frac{1}{2}}$.

Arithmetic Novelty 60

Problem: Find the fourth power of $\sqrt{1 + \sqrt{1 + \sqrt{1}}}$.

Solution: Let $x = \sqrt{1 + \sqrt{1 + \sqrt{1}}}$. Since $\sqrt{1} = 1$,

$$x = \sqrt{1 + \sqrt{2}},$$

then

$$x^2 = 1 + \sqrt{2} \text{ and } x^4 = \left(x^2\right)^2 = 1 + 2\sqrt{2} + 2 = 3 + 2\sqrt{2}.$$

Arithmetic Novelty 61

Problem: A positive number is mistakenly divided by six instead of being multiplied by six. Based on the correct answer, what is the error thus committed to the nearest percent.

Solution: Let N represent the positive number, so that the incorrect result is $\frac{N}{6}$. Since the correct product is $6N$, the error is $6N - \frac{N}{6} = \frac{35N}{6}$. Therefore, the percent error based on the correct result is

$$\frac{\text{error}}{\text{correct result}} \cdot 100 = \frac{3{,}500}{36} \approx 97\%.$$

Arithmetic Novelty 62

Problem: An operation represented by $*$ is defined as $a*b = \frac{ab}{a+b}$. The challenge here is to find $4 * (4 * 4)$.

Solution: Since $a*b = \frac{ab}{a+b}$, we have $4*4 = \frac{4 \cdot 4}{4+4} = \frac{16}{8} = 2$. Therefore,

$$4 * (4 * 4) = 4 * 2 = \frac{4 \cdot 2}{4+4} = \frac{8}{6} = \frac{4}{3}.$$

Arithmetic Novelty 63

Problem: Find the simplified value of $\left(-\frac{1}{125}\right)^{-\frac{2}{3}}$.

Solution: $\left(-\frac{1}{125}\right)^{-\frac{2}{3}} = (-125)^{\frac{2}{3}} = \left[(-125)^{\frac{1}{3}}\right]^2 = (-5)^2 = 25.$

Arithmetic Novelty 64

Problem: Find the sum of the numbers from 1 to 100 without adding the numbers consecutively.

Solution: A simple technique was popularized by the famous German mathematician Carl Friedrich Gauss, who in elementary school was given the task of adding the numbers from 1 to 100 so that the teacher could keep the class busy. He solved the problem within one minute in the following way. Rather than add the numbers consecutively, as most of his classmates did, he added them in pairs as follows: $1 + 100 = 101$, $2 + 99 = 101$, $3 + 98 = 101$ and so on. He realized that there were 50 such pairs, so that the total sum of the 100 numbers would be $50 \cdot 101 = 5,050$.

Arithmetic Novelty 65

Problem: Calculate the average speed for a round trip with a "going" average speed of 30 miles per hour and a "returning" average speed of 60 miles per hour.

Solution: One would think that their average speed for the entire trip is 45 miles per hour (calculated as $\frac{30+60}{2} = 45$). This is the wrong answer. The two speeds were achieved for different lengths of time and therefore cannot get the same weight. The outgoing trip at 30 mph took twice as long as the return trip of 60 mph, and therefore ought to get twice the weight in the calculation of the average round-trip speed. This would then bring the calculation to the following: $\frac{30+30+60}{3} = 40$ mph, which is the correct average speed.

For those not convinced by this argument, try something a bit closer to "home." A question can be posed about the grade a student deserves who scored 100% on nine of ten tests in a semester and on one test scored only 50%. Would it be fair to assume that this student's performance for the term was 75% (i.e., $\frac{100+50}{2}$)? This suggestion will lead toward applying appropriate weight to the two scores in consideration. The 100% was achieved nine times as often as the 50% and therefore ought to get the appropriate weight. Thus, a proper calculation of the student's average ought to be $\frac{9(100)+50}{10} = 95$. This is clearly more just!

An astute reader may now ask, "What happens if the rates to be averaged are not multiples of one another?" For the speed problem above, one could find the time "going" and the time "returning" to get the total time, and then with the total distance calculate the total rate, which is, in fact, the average rate.

There is a more efficient way to solve this kind of problem by introducing a concept called the *harmonic mean,* which is the mean of a harmonic sequence. The name *harmonic* may come from the fact that one such harmonic sequence is $\frac{1}{2}, \frac{1}{3}, \frac{1}{4}, \frac{1}{5}, \frac{1}{6}, \frac{1}{7}, \frac{1}{8}$, and if one takes guitar string of these relative lengths and strums them together a harmonious sound will result.

This frequently misunderstood mean (or average) usually causes confusion, but to avoid this, once we identify that we are to find the average of rates (i.e., the harmonic mean), then we have an efficient formula for calculating the harmonic mean for rates over the same base. In the above situation, the rates were for the same distance (round trip legs). The harmonic mean for two rates, a and b, is $\frac{2ab}{a+b}$. For three rates, a, b, and c, the harmonic mean is $\frac{3abc}{ab+bc+ac}$. You can see the pattern evolving, so that for four rates the harmonic mean is $\frac{4abcd}{abc+abd+acd+bcd}$. Applying this to the above speed problem gives us: $\frac{2 \cdot 30 \cdot 60}{30+60} = \frac{3,600}{90} = 40$. Now let's consider the related next problem.

Arithmetic Novelty 66

Problem: On Monday, a plane makes a round-trip flight from New York City to Washington with an average speed of 300 miles per hour. On Tuesday, there is a wind of constant speed (50 miles per hour) and direction (blowing from New York City to Washington). With the same speed setting as on Monday, this same plane makes the same round trip on Tuesday. Will the Tuesday trip require more time, less time, or the same time as the Monday trip?

Solution: The two speeds of the "windy trip" cannot be weighted equally as each leg lasts different lengths of time. Therefore, the duration of each leg should be calculated and then appropriately apportioned to the related speeds. We can use the harmonic mean formula (see Arithmetic Novelty 65) to find the average speed for the "windy trip."

The harmonic mean is $\frac{(2)(350)(250)}{250+350} = 291.667$, which is slower than the no-wind trip. This topic is not only useful, but also sensitizes the reader to the notion of weighted averages, a very important concept to remember.

Arithmetic Novelty 67

Problem: The three-digit number $2a3$ is added to the number 326 to result in the three-digit number $5b9$. If the number $5b9$ is divisible by 9, then what is the sum of $a + b$ equal to?

Solution: The key factor that can simplify the solution to this problem is that a number is divisible by 9 if and only if the sum of its digits is divisible by 9. Therefore, the sum of the digits $(5 + b + 9)$ must be a multiple of 9, so that the digit $b = 4$. We also know that $2a3 = 5b9 - 326 = 549 - 326 = 223$, therefore, $a = 2$ and $a + b = 6$

Arithmetic Novelty 68

Problem: If $S = 1! + 2! + 3! + 4! + \cdots + 98! + 99!$, then we seek the units digit of the value of S. (Note: the symbol $n!$ denotes $1 \cdot 2 \cdot 3 \cdot 4 \cdots (n-1) \cdot n$).

Solution: Beginning with

$$S = 1! + 2! + 3! + 4! \cdots + 98! + 99!$$
$$= 1 + 2 + 2 \cdot 3 + 2 \cdot 3 \cdot 4 + 2 \cdot 3 \cdot 4 \cdot 5 + \cdots 99!$$
$$S = 1 + 2 + 6 + 24 + 120 + \cdots + 10k,$$

where k is a positive integer, and indicates that the last number is a multiple of 10. In other words, each of the numbers in the set 5!, 6!, $\cdots$, 99! contains the factors of 2 and 5 and thus is a multiple of 10. The sum of the units digits of S is therefore $1 + 2 + 6 + 4 = 13$. Thus, the sought after units digit is 3, since all the other numbers have 0 as a last digit.

Arithmetic Novelty 69

Problem: Which is greater: 25% of $76, or 76% of $25?

Solution: The answer is that they are both the same: $19. In this case, what appears to be a curious problem should not be bogged down with the calculation. Rather, we should notice that there is a commutativity here, since in both cases we are multiplying the "same two numbers."

In one case we have $\frac{25}{100} \cdot \$76 = \frac{25 \cdot 76}{100}$ (= $19), and the other we find that $\frac{76}{100} \cdot \$25 = \frac{76 \cdot 25}{100}$ (= $19).

Arithmetic Novelty 70

Problem: Find the units digit for the following sum: $13^{25} + 4^{81} + 5^{411}$.

Solution: Examine the patterns that exist in the powers of three different sets of numbers. This will help familiarize us with the cyclical pattern for the final digits of the powers of numbers.

For powers of 13, we obtain:

$$13^1 = \quad 1\underline{3} \qquad 13^5 = \quad 37129\underline{3}$$
$$13^2 = \quad 16\underline{9} \qquad 13^6 = \quad 482680\underline{9}$$
$$13^3 = \quad 219\underline{7} \qquad 13^7 = \quad 6274851\underline{7}$$
$$13^4 = 2856\underline{1} \qquad 13^8 = 72573172\underline{1}$$

The units digits for powers of 13 repeat as 3, 9, 7, 1, 3, 9, 7, 1, . . . in cycles of 4. Thus, 13^{25} has the same units digit as 13^1, or 3.

For powers of 4, we obtain:

$$4^1 = \quad 4 \qquad 4^5 = \quad 1{,}024$$

$$4^2 = \quad 16 \qquad 4^6 = \quad 4{,}096$$

$$4^3 = \quad 64 \qquad 4^7 = 16{,}384$$

$$4^4 = 256 \qquad 4^8 = 65{,}536$$

The units digits for powers of 4 repeat as 4, 6, 4, 6, 4, 6, . . . in cycles of 2. Thus, 4^{81} has the same units digit as 4^1, which is 4. The units digit for powers of 5 must be 5 (i.e., 5, 25, 125, 625, etc.). The sum we are looking for is $3 + 4 + 5 = 12$, which has a units digit of 2.

Arithmetic Novelty 71

Problem: Express in simplest form the value of the following product:

$$\left(1+\frac{1}{1}\right)\left(1+\frac{1}{2}\right)\left(1+\frac{1}{3}\right)\left(1+\frac{1}{4}\right)\cdots\left(1+\frac{1}{n}\right)\cdots\left(1+\frac{1}{100}\right).$$

Solution: When we calculate the value of each parenthetical expression, we get the following: $\frac{2}{1} \cdot \frac{3}{2} \cdot \frac{4}{3} \cdot \frac{5}{4} \cdot \quad \cdots \quad \cdot \frac{100}{99} \cdot \frac{101}{100} = \frac{101}{1} = 101$, which results when we cancel consecutive numerators and denominators.

Arithmetic Novelty 72

Problem: Without using division, determine whether 9 is a factor of each of these numbers:

32,471,921
9,652,689
8,126,53,789.

Solution: In the base-10 system, there is a very clever way of determining whether 3 or 9 is a factor of the number by taking the sum of the number's digits. If the sum of the digits is divisible by 3, then the number is divisible by 3, and if the sum of the digits is divisible by 9, then the number is divisible by 9. We now consider the sum of the digits of each of the numbers to see if they are divisible by 9:

- $3 + 2 + 4 + 7 + 1 + 9 + 2 + 1 = 29$, which is not divisible by 3 or 9.
- $9 + 6 + 5 + 2 + 6 + 8 + 9 = 45$, which is divisible by both 3 and 9.
- $8 + 1 + 2 + 6 + 5 + 3 + 7 + 8 + 9 = 48$, which is divisible by 3 but not divisible by 9.

By the way, if the first digit sum is too large to determine visually whether it is divisible by 3 or 9, you can continue the process and take the sum of the digits of that sum and proceed as earlier. For example, in the last two number sums were 45 and 48. The digit sum of 45 is $4 + 5 = 9$, which is divisible by 3 and 9, and the digit sum of 48 is $4 + 8 = 12$, which is divisible by 3 and not by 9.

Arithmetic Novelty 73

Problem: Simplify each of the following without a calculator:

a) $\dfrac{729^{35} - 81^{52}}{27^{69}}$,

b) $\dfrac{6 \cdot 27^{12} + 2 \cdot 81^{9}}{8000000^{2}} \cdot \dfrac{80 \cdot 32^{3} \cdot 125^{4}}{9^{19} - 729^{6}}$.

Solution: Although one may be tempted to use a calculator to evaluate this expression, all too often our expectations for the calculator are too high, and the result comes back with an "error" message. With the knowledge of powers, the problem can be rather cleverly evaluated as follows:

a) $\dfrac{729^{35} - 81^{52}}{27^{69}} = \dfrac{(3^{6})^{35} - (3^{4})^{52}}{(3^{3})^{69}} = \dfrac{3^{210} - 3^{208}}{3^{207}} = \dfrac{3^{208} \cdot (3^{2} - 1)}{3^{207}} = 3 \cdot 8 = 24.$

This expression can be simplified by breaking up the numbers into prime factors as follows:

b)
$$\frac{6 \cdot 27^{12}+2 \cdot 81^{9}}{8000000^{2}} \cdot \frac{80 \cdot 32^{3} \cdot 125^{4}}{9^{19}-729^{6}} = \frac{2 \cdot 3 \cdot (3^{3})^{12}+2 \cdot (3^{4})^{9}}{(3^{2})^{19}-(3^{6})^{6}} \cdot \frac{2^{4} \cdot 5 \cdot (2^{5})^{3} \cdot (5^{3})^{4}}{(2^{3} \cdot 2^{6} \cdot 5^{6})^{2}}$$

$$=\frac{2 \cdot 3^{37}+2 \cdot 3^{36}}{3^{38}-3^{36}} \cdot \frac{2^{19} \cdot 5^{13}}{2^{18} \cdot 5^{12}} = \frac{2 \cdot 3^{36}(3+1)}{3^{36}(3^{2}-1)} \cdot 2 \cdot 5 = \frac{2(3+1)}{3^{2}-1} \cdot 2 \cdot 5 = 10.$$

Arithmetic Novelty 74

Problem: We are given the following four numbers

7,895

13,127

51,873

7,356.

What percent of their sum is their average?

Solution: When faced with this question, we typically do exactly what is being called for. That is, first we find the sum of the four numbers; then we find their average. Finally, we divide and convert to required percent: $7{,}895 + 13{,}127 + 51{,}873 + 7{,}356 = 80{,}251$ and $\frac{80{,}251}{4}=$ $20{,}062.75$ and $\frac{20{,}062.75}{80251}=\frac{1}{4}=0.25$, i.e., 25 %.

However, it might be wise, and perhaps simpler, to first consider the general case. We shall let the sum of these numbers be represented by S. Then their average is $\frac{S}{4}$. To find what percent the average is of the sum, we first divide $\frac{\frac{S}{4}}{S}=\frac{1}{4}$. We now change $\frac{1}{4}$ to a percent – which the problem calls for – to get 25%. We have avoided a great deal of unnecessary calculation by simply stepping back from the problem and considering the general case. Observe how it presents us with the answer. What may be surprising is that the result is independent of the four given numbers!

Chapter 2

Logic Novelties

The concept of logic plays an important role in mathematics. Logical thinking often leads to some clever problem-solving techniques. Much more time should be devoted during the school years to this aspect of mathematics to encourage favorable feeling for the subject. In this section we provide a plethora of problems that rely largely on logical thinking. A new appreciation for problem solving through logical thinking should evolve.

Logic Novelty 1

Problem: In a battle, 5% of the army was killed and $12\frac{1}{2}\%$ of the remainder wounded. Furthermore, $16\frac{2}{3}\%$ of the wounded later died, that is, were mortally wounded. If there were 290 more soldiers killed than were mortally wounded, how many soldiers were there in the army?

Solution: Let x = the number of soldiers in the army. The number of soldiers in the army who were killed is $\frac{1}{20}x$. Furthermore, $\frac{19}{20}x$ = the number of soldiers in the army who were not killed. Since $12\frac{1}{2}\% = \frac{1}{8}$, we then have $\left(\frac{1}{8}\right)\cdot\left(\frac{19}{20}x\right)$ as the number of soldiers in the army who were wounded. Again since $16\frac{2}{3}\% = \frac{1}{6}$, we have $\left(\frac{1}{6}\right)\cdot\left(\frac{1}{8}\right)\cdot\left(\frac{19}{20}x\right)$ = the number of soldiers in the army who were mortally wounded. We then have the number of soldiers killed as $\frac{1}{20}x = \left(\frac{1}{6}\right)\cdot\left(\frac{1}{8}\right)\cdot\left(\frac{19}{20}x\right)+290$.

When we multiply both sides of the equation by the lowest common denominator (i.e., $6 \cdot 8 \cdot 20$), we get $(8)(6)x = 19x + (6)(8)(20)(290)$, and thus, $x = 9{,}600$, which was the number of soldiers in the army.

Logic Novelty 2

Problem: There were five candidates, A, B, C, D, and E, in a class election. Candidate A was elected with 26 votes. Candidate B placed second, and candidate E came in last with 2 votes. If no two candidates had the same number of votes, and there were 45 votes cast in total, what is the smallest number of votes that B could have received?

Solution: Since the total number of votes is 45, and candidates A and E together have 28 votes, the votes for $B + C + D = 17$. If B has 6 votes, then C or D (not both) would have at least 6 votes. However, B must have more votes than C or D. Therefore, B must have more than 6 votes. Thus, the smallest number of votes that B could have received is 7.

Logic Novelty 3

Problem: How many minutes after two o'clock will the minute hand of a clock overtake the hour hand?

Solution: Assume that the hour hand remains stationary. The minute hand would then have to travel 10 minutes to the "2" to overtake the hour hand. To account for the actual movement of the hour hand, multiply 10 minutes by the "magic fraction,"[1] $\frac{12}{11}$ to get $10 \cdot \left(\frac{12}{11}\right) = \frac{120}{11} = 10\frac{10}{11}$. Hence, the minute hand will overtake the hour hand at $2{:}10\frac{10}{11}$, which is in $10\frac{10}{11}$ minutes.

---

[1] See Arithmetic Novelty 36.

Logic Novelty 4

Problem: Sam and Jack together have 14 marbles. Jack and Max together have 10 marbles. Sam and Max together have 12 marbles. What is the maximum number of marbles that any one of these boys can have?

Solution: Let the initial letter of each boy's name represent the number of marbles he has.

Therefore:

$$S + J = 14, \qquad \text{(a)}$$

$$J + M = 10, \qquad \text{(b)}$$

$$S + M = 12. \qquad \text{(c)}$$

Subtract (b) from (c) to get:

$$S - J = 2, \qquad \text{(d)}$$

$$S + J = 14. \qquad \text{(a)}$$

Add (d) and (a) to get $S = 8$. It then follows (by substitution) that $J = 6$, and $M = 4$. The maximum number of marbles is 8.

Logic Novelty 5

Problem: What is the maximum number of points of intersection of five equal-sized circles?

Solution: When two circles intersect, there are at most two points of intersection. When a third circle intersects each of the first two, it meets them in at most 2 points each, or 4 additional points (i.e., three circles intersect in at most 6 points). A fourth circle meets each of the previous three circles at most twice, for 6 additional points (i.e., four circles meet in at most 12 points). A fifth circle meets each of the previous four circles at most twice, for an additional 8 points. Therefore, five circles intersect in at most 20 points. Figure 2-1 is a summary of the above explanation.

Number of circles	Maximum number of points of intersection
2	2
3	$2 + 2(2) = 6$
4	$6 + 2(3) = 12$
5	$12 + 2(4) = 20$

Figure 2-1

Logic Novelty 6

Problem: Lines are drawn across a rectangle parallel to the sides, as in Figure 2-2. How many rectangles can be counted?

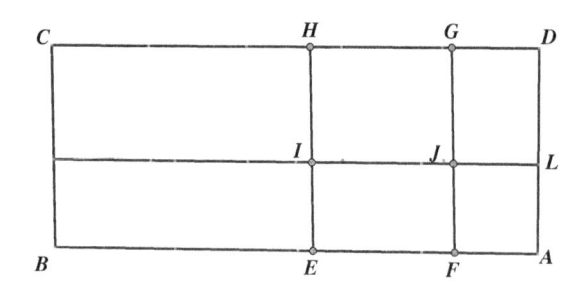

Figure 2-2

Solution: The following is a list of all 18 rectangles seen in Figure 2-2:

AEIL	*LIHD*	*AEHD*
AFJL	*LJGD*	*EFGH*
ABKL	*LKCD*	*FBCG*
EIJF	*IHGJ*	*ABCD*
EIKB	*IHCK*	*AFGD*
FJKB	*JGCK*	*EBCH*

Logic Novelty 7

Problem: Find the difference between the maximum number of school days in a 31-day month and the minimum number of school days in a 30-day month with no holidays. (Assume a five-day school week.)

Solution: Out of the first 28 days in any month, 20 are school days. A 31-day month can have a maximum of 23 school days, and a 30-day month can have a minimum of 20 school days if the two "extra" days comprise a weekend. This provides a difference of three days.

Logic Novelty 8

Problem: A slab rests on rollers (Figure 2-3), each 1 foot in diameter. Assuming that there is no slippage, how far forward does the slab move when the rollers have made one revolution?

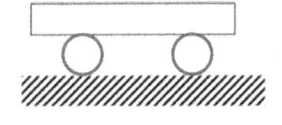

Figure 2-3

Solution: Since, in one revolution, the slab moves π feet relative to the roller while the roller moves π feet relative to the ground, the slab moves 2π feet relative to the ground.

Logic Novelty 9

Problem: Two men are sorting out a financial question. Al owes Harvey 4 dollars and Harvey owes Al 5 dollars. Then Al gives Harvey 6 dollars. What then needs to be done to cancel the debt?

Solution: When Al gives Harvey 6 dollars, then Harvey owes Al 6 dollars more than before, that is he owes him 11 dollars. But since Al owes Harvey 4 dollars, Harvey then only owes Al 7 dollars. To cancel the debt, Harvey must give Al 7 dollars.

Logic Novelty 10

Problem: What is the least number of dots which may be moved to make the triangular arrangements of dots shown in Figure 2-4 to reverse its direction?

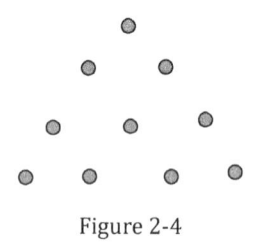

Figure 2-4

Solution: In Figure 2-5, we show that only three dots need to be moved to change direction of the triangle.

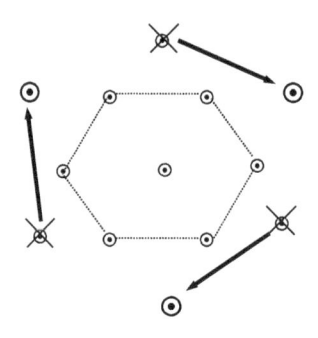

Figure 2-5

Logic Novelty 11

Problem: Joshua was three times as old as his brother two years ago and five times as old as his brother two years before that. In how many years will the ratio be 2:1?

Solution: Let x = the age of Joshua's brother four years ago. Then $5x + 2 = (x + 2)$, so that $x = 2$. Therefore, Joshua is 14 years old, and his brother is 6 years old. Thus, in two years the ratio will be 16:8, or 2:1.

Logic Novelty 12

Problem: What is the measure of an angle formed by the hour hand and the minute hand of a clock at 2:15?

Solution: At 2:15, the minute hand is exactly on the "3." At the same time, the hour hand has traveled $\frac{1}{4}$ the distance between the "2" and the "3," or 30°. Therefore, the angle is $\frac{3}{4}$ of this interval, or $\frac{3}{4} \cdot 30° = 22\frac{1}{2}°$.

Logic Novelty 13

Problem: Ernie can paint a room in 36 hours. After he had worked for 20 hours, Alice joined him and they finished the wall in 10 hours more. If they had done the job together the entire time how long would it have taken them to finish the job?

Solution: After Ernie works for 20 hours, he completes $\frac{20}{36} = \frac{5}{9}$ of the job. Since the remaining $\frac{4}{9}$ of the job is done by both working together for 10 hours. If x is the number of hours in which both working together can complete the entire job, then $\frac{4}{9} = \frac{10}{x}$, and thus, $x = 22\frac{1}{2}$ hours.

Logic Novelty 14

Problem: If a pipe 3 inches in diameter empties a tank of water in 4 minutes, how long will it take a pipe 2 inches in diameter to empty the same tank?

Solution: The circular cross-section of the pipe is the critical aspect of the pipe's size. Since the ratio of the diameters is $\frac{3}{2}$, the ratio of their circular cross-sections (circle areas) is $\frac{3^2}{2^2} = \frac{9}{4}$. Since the area of the cross-section of each pipe varies inversely as the time it requires to empty the tank, we set up the following proportion $\frac{9}{4} = \frac{x \text{ minutes}}{4 \text{ minutes}}$; therefore, $x = 9$, which is the time that the smaller pipe will require to empty the tank.

Logic Novelty 15

Problem: Find the next number in the following sequence: 2, 9, 23, 44, 72, ___.

Solution: Consider the differences between the consecutive terms, which are 7, 14, 21, 28. These are consecutive multiples of 7. Therefore, to remain consistent, the next difference should be 35, which when added to the last term yields 107.

Logic Novelty 16

Problem: A farmer plants corn in a square grid so that the number of rows and columns are equal. He increases the size of his corn field equally in the number of rows and columns to create a new field which contains 211 additional corn plants. How many corn plants were in one row of the original field?

Solution: If the original field had x corn plants in a row, then the number of corn plants would be x^2. If the new field has h additional plants in each row, then it will contain $(x + h)^2$ plants. Thus,

$$x^2 + 211 = (x + h)^2$$
$$x^2 + 211 = x^2 + 2hx + h^2$$
$$211 = h^2 + 2hx$$
$$211 = h(h + 2x).$$

This appears to be rather complex, since it is a quadratic equation in h, but also contains x. So, let's consider the number 211, which is a prime number. Since h and x must be whole numbers, h must be l, and $h + x$ must be 211. Therefore, $2x = 210$ and $x = 105$.

Logic Novelty 17

Problem: David and Lisa are playing a game. Whoever loses a match gives the other person a penny. When they were done, David had won 3 times, but Lisa had 8 more pennies than she had started with. How many times did David and Lisa play the game?

Solution: Since David won 3 times, Lisa must also have won 3 times just to get back to where she began. Then she won 8 more times. Thus, she must have won a total of 11 matches and lost 3, which means that they played 14 matches.

Logic Novelty 18

Problem: On a shelf in Lara's basement, there are 3 boxes. One contains only nickels, one contains only dimes, and one contains a mixture of nickels and dimes. The three labels, "Nickels," "Dimes," and "Mixed" fell off, and were all put back on the wrong boxes. Without looking, Lara can select one coin from one of the mislabeled boxes and then correctly label all 3 boxes. From which box should Lara select the coin?

Solution: You may reason that the "symmetry" of the problem dictates that whatever we can say about the box mislabeled "Nickels" could just as well have been said about the box mislabeled "Dimes." Thus, if Lara chooses a coin from either of these boxes, the results would be the same. This eliminates selecting the box labeled Nickels and the box labeled Dimes. One should, therefore, concentrate the investigation on what happens if one chooses a coin from the box mislabeled "Mixed." Suppose Lara selects a nickel from the Mixed box. Since this box is mislabeled, it cannot be the mixed box and must be the Nickels box. Since the box marked Dimes cannot really be dimes, it must be the Mixed box. This means that the third box, marked Nickels, is the Dimes box. The correct answer is for Lara to select the coin from box mislabeled "Mixed."

Logic Novelty 19

Problem: Max and Sam are about to have cupcakes as a snack. Max has 3 cupcakes, while Sam has 5 cupcakes. Just as they are about to eat, Jack comes along. Jack has no cupcakes, but he does have 40 cents to pay for his share. The cupcakes are divided equally among Max, Sam and Jack. When Jack leaves, Max and Sam divide up the 40 cents. How much does Sam receive?

Solution: At first glance, one might decide to divide the 40 cents in the ratio of 3:5, the number of cupcakes each boy contributed. However, this neglects the fact that Max and Sam each ate $\frac{1}{3}$ of the cupcakes. Let's organize the data carefully, as in Figure 2-6. Eight cupcakes are placed into the common "pot" to start.

	Max	**Sam**	**Jack**
Amount put into the "pot"	3	5	40 cents
Amount eaten from "pot"	$\dfrac{8}{3}$	$\dfrac{8}{3}$	$\dfrac{8}{3}$
Amount left in "pot"	$\dfrac{9}{3} - \dfrac{8}{3} = \dfrac{1}{3}$	$\dfrac{15}{3} - \dfrac{8}{3} = \dfrac{7}{3}$	40 cents

Figure 2-6

Thus, Max gets $\frac{1}{3}$ of the money and Sam gets $\frac{7}{3}$, for a 1:7 ratio. Max receives 5 cents, and Sam receives 35 cents.

Logic Novelty 20

Problem: Consider a specific job involving three people, whom we will call x, y, and z. The workers x and y working together can do the job in 10 hours, the workers x and z working together can do the job in 15 hours, and the workers y and z working together can do the job together and 18 hours. How many hours would it take to do the entire job when all three people are working together?

Solution: Let x, y, and z represent the amount of the job that x, y, and z can do in one hour. We then have the following:

$$x + y = \frac{1}{10} = \frac{9}{90},$$

$$x + z = \frac{1}{15} = \frac{6}{90},$$

$$y + z = \frac{1}{18} = \frac{5}{90}.$$

By adding these three parts we will show that together they will have accomplished $2(x + y + z) = \frac{20}{90} = \frac{2}{9}$, or $x + y + z + = \frac{1}{9}$ of the job working together. Thus, it will take them 9 hours to complete the job working together.

Logic Novelty 21

Problem: A master tailor can do half a job in one hour. He has several apprentices, each of whom would do $\frac{1}{6}$ of the job in an hour. How long would it take the master tailor working with two apprentices to finish the job?

Solution: In one hour, the master tailor would finish $\frac{1}{2}$ of the job. Since each apprentice would do $\frac{1}{6}$ of the job in one hour, together they would accomplish $\frac{1}{3}$ of the job. When all three tailors are working together, they will accomplish in one hour $\frac{1}{2} + \frac{1}{3} = \frac{5}{6}$ of the job. Therefore, working together, they will require $\frac{6}{5} = 1\frac{1}{5}$ hours to do the whole job.

Logic Novelty 22

Problem: There are two equal size half-bottles of wine: one contains red wine, and the other one contains white wine. If one takes a spoonful of the red wine and pours it into the bottle of white wine and then after mixing pours the same spoonful of the mixture back into the red wine bottle, is there more red wine in the white-wine bottle, or more white wine in the red-wine bottle?

Solution: Typically, this problem causes great confusion unless one is clever enough to use the following procedure. Since the size of the spoonful was not given but the same spoon was used in both transports, there is no reason why one cannot consider a very large spoon. One could even use a spoon that is big enough to carry the entire volume of the red wine bottle. Therefore, when the red wine is poured into the white-wine bottle the resulting mixture is 50% white wine and 50% red wine. When this mixture is then moved from the white-wine bottle with the spoon that holds half the volume of the bottle and poured into the empty red-wine bottle, one finds that there is as much red wine in the white-wine bottle as there is white wine in the red-wine bottle. This answers the question.

Logic Novelty 23

Problem: A local movie theater is showing two short films, one in each of two adjoining theaters. The first film runs for 36 minutes; the second runs for 45 minutes. They both begin at 8:45 a.m., and run continuously until 5:45 p.m. At what times during the day will both films again start at exactly the same time?

Solution: The two films run for 36 minutes and 45 minutes and we need to find the common multiples of both runtimes. For 36, two factors are 9 and 4, while two factors of 45 are 9 and 5. Thus, the least common multiple for both runtimes is $9 \cdot 4 \cdot 5 = 180$, which indicates that exactly every 3 hours both films will have been completed. Therefore, the films start together at 11:45 a.m. and 2:45 p.m.

Logic Novelty 24

Problem: There are 12 trains that pass through the towns of Westwood and Hillsdale. One-half of these trains stop at Westwood, one-third of these trains stop at Hillsdale, and one-fourth of these trains stop at both Westwood and Hillsdale. How many of these trains stop at neither of these towns?

Solution: Let's solve this problem with a Venn diagram, as shown in Figure 2-7. The large circle contains 12 trains. If half of the trains stop at Westwood, the Westwood circle must contain 6, shown as 3+3. If one-third of the trains stop at Hillsdale, then that circle must contain 4, which is shown as 3+1. When one quarter of the trains stop at both towns, the intersection of the two circles must contain 3. Since there were 12 trains traveling and we have seven making stops, there must be 5 that have made no stops. Using a Venn diagram certainly simplifies the solution.

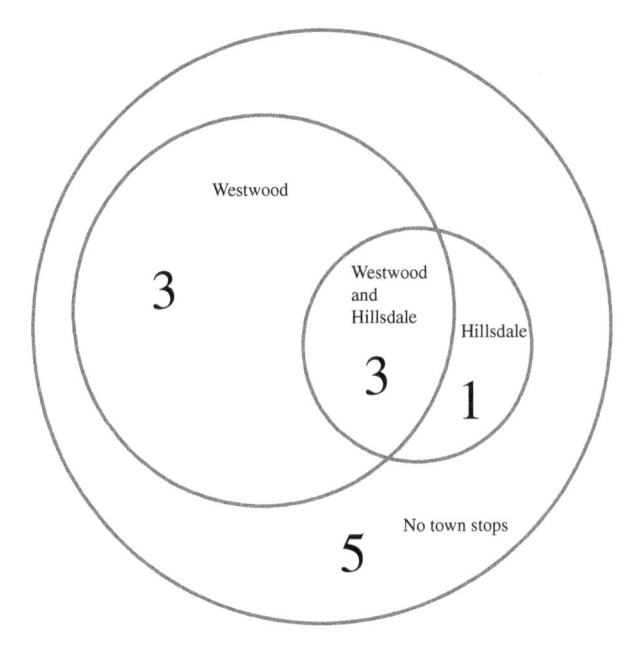

Figure 2-7

Logic Novelty 25

Problem: It takes 7 seconds for a clock to strike 7 "bongs" at 7:00 o'clock. At this same rate, how long will it take for the clock to strike 10 o'clock? (Assume that the actual "bongs" take no time at all.)

Solution: We see that there are 6 intervals. Thus, each interval is $7 \div 6$, or $\frac{7}{6}$ seconds long. When the clock strikes 10:00 o'clock there are 9 intervals. Since each interval is $\frac{7}{6}$ of a second long, it takes the clock $9 \cdot \frac{7}{6} = \frac{21}{2} = 10\frac{1}{2}$ seconds.

Logic Novelty 26

Problem: A large auditorium requires painting. If it takes 10 people 6 days to complete the job, how long will it take 7 people to do the same job, when all work at the same pace?

Solution: We can consider an extreme case, where one person does this job and it takes that person 60 days. When 7 people do a job that takes 60 days for one person, then it would be completed in $\frac{60}{7} = 8\frac{4}{7} \approx 8.57$ days.

Logic Novelty 27

Problem: A rectangle's sides measure 10 units and 5 units. A second rectangle, with sides measuring 6 units and 4 units, overlaps the first rectangle, as shown in Figure 2-8. What is the difference between the two non-overlapping regions of the two rectangles?

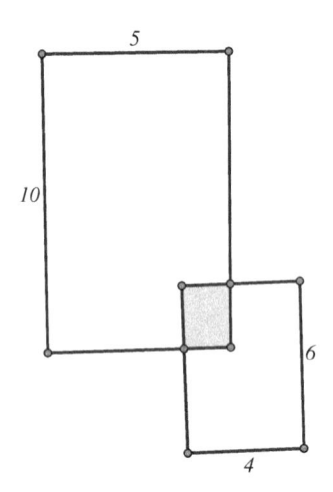

Figure 2-8

Solution: If one attempts to attack this problem directly, they will quickly find themselves at a loss. Instead, let's consider the drawing from a different point of view; that is, since the region of overlap is not determined or specified, let us suppose the rectangles are attached only at a single vertex, with a region of area 0 overlapping, as shown in Figure 2-9.

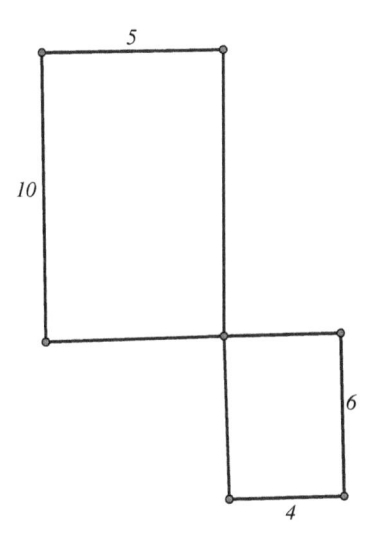

Figure 2-9

Now we see that the difference of the areas of the two rectangles is $50 - 24 = 26$ square units. Now consider the rectangles so that there will be a single unit of overlap as shown in Figure 2-10, where the difference of the non-overlapping areas is $49 - 23 = 26$ square units.

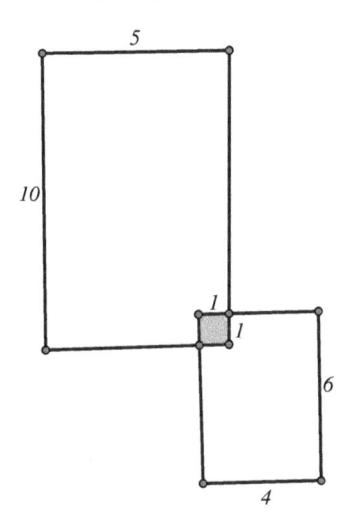

Figure 2-10

Finally, let's enclose the smaller rectangle entirely within the large one, as shown in Figure 2-11. Again, the non-overlapping region is 26 square units. Once more, we emphasize the value of considering a problem from a different point of view.

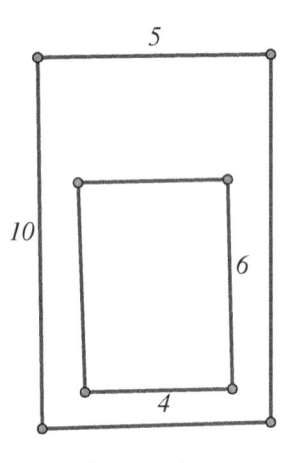

Figure 2-11

Notice that we need not consider each of the situations described above. For some, the initial idea where the rectangles merely share a common point might have been sufficient. For others, however, the subsequent situations might be necessary for them to conclude the correct answer.

Logic Novelty 28

Problem: A basketball tournament has 36 teams. One loss eliminates a team. How many games must be played to get a tournament winner?

Solution: Typically, one simulates a tournament beginning with the first eighteen games and successively eliminating teams until only one remains. This could take quite some time and most likely lead to confusion when dealing with large numbers. A much more efficient method would be to look at the end result: the winner. Working backwards, one may ask how many losers must there have been? There must be 35 losers. How do we get to 35 losers? There must be 35 games! And so, the problem is solved.

Logic Novelty 29

Problem: How many angles are formed by 10 distinct rays, shown in Figure 2-12, with a common endpoint?

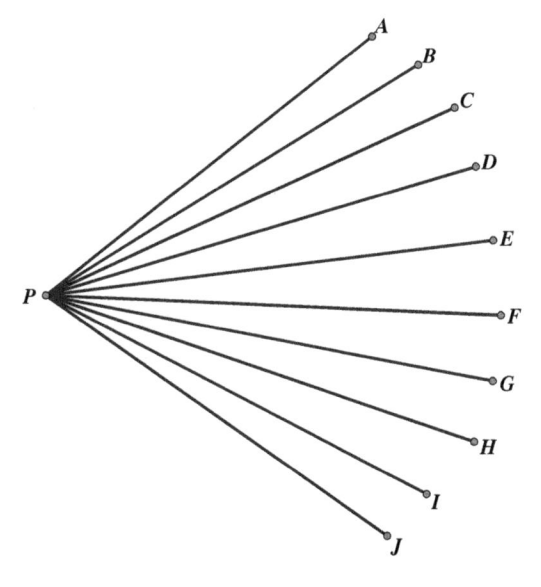

Figure 2-12

Solution: One can draw a large diagram as shown in Figure 2-12 to answer this question. One could start counting the smaller angles, such as ∠APB, ∠BPC, ∠CPD, and then larger angles, such as ∠APC, ∠BPD, and even larger angles, ∠APE and so on. However, the drawing soon becomes rather confusing as one can lose track of the angles being counted. Let's start with a simpler case and see if a pattern emerges as we increase the number of rays. This data is summarized in Figure 2-13.

Number of rays	1	2	3	4	5	6	7	8
Number of angles formed	0	1	3	6	10	15	21	28

Figure 2-13

A quick review of the number of angles formed as the number of rays increases by 1 reveals a pattern: 0, 1, 3, 6, 10, etc. We have a sequence whose differences form a simple arithmetic progression: 1, 2, 3, 4, 5, etc. Continuing this sequence to 10 terms is simple: 0, 1, 3, 6, 10, 15, 21, 28, 36, 45. Therefore, with 10 rays there will be 45 different angles.

Note: These numbers are referred to as the "triangular" numbers because they can be represented geometrically in a triangular array, as can be seen in Figure 2-14.

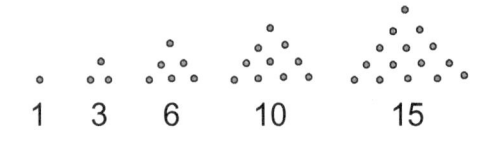

Figure 2-14

Logic Novelty 30

Problem: Two fixed points, x and y, are located 100 feet apart. A movable point P moves from x to y in the following uniform manner: During the first half of each minute P moves 3 feet toward y, and during the second half of each minute it returns 2 feet back toward x. In how many minutes will P reach y?

Solution: It should be clear that after 1 complete minute, the point P has progressed 1 foot toward y. After 2 complete minutes it has progressed 2 feet toward y. After 10 complete minutes it has progressed

10 feet toward y. By this reasoning, we find that after 97 complete minutes, point P has progressed 97 feet toward y. However, during the first half of the 98th minute, P will travel 3 feet beyond the 97-foot mark. In other words, P will reach the 100-foot mark in that time interval. Therefore, point P will have traveled 100 feet in $97\frac{1}{2}$ minutes.

Logic Novelty 31

Problem: Nancy has 5 blue ribbons, 3 red ribbons, and 7 yellow ribbons in a drawer. If Nancy is blindfolded, what is the least number of ribbons she must take from the drawer to be sure to get two ribbons of the same color?

Solution: We must assume the worst possible luck. On her first three selections, Nancy could take ribbons of 3 different colors. However, she cannot choose a different color on her fourth selection, since she already has one of each color. Hence, she is assured of matching one of her previous selections on the fourth draw.

Logic Novelty 32

Problem: Evelyn has 5 blue ribbons, 3 red ribbons, and 7 yellow ribbons in a drawer. If she is blindfolded, what is the least number of ribbons she must take from the drawer to be sure to get one of each color?

Solution: Again, we must assume that Evelyn has the worst possible luck. The worst that can happen is that she takes 7 yellow ribbons and 5 blue ribbons for her first 12 selections. On her next selection, she must get a red ribbon, as there are no longer any blue or yellow ribbons left. Therefore, Evelyn is assured of getting one ribbon of each color on her thirteenth selection.

Logic Novelty 33

Problem: Carol took a 20-question multiple choice test last Thursday. The test was scored +5 for each correct answer, −2 for an incorrect answer, and 0 if the question was not answered. Carol scored 44, even though she omitted some of the questions. How many questions did Carol omit?

Solution: We first reason that Carol must have answered at least 10 questions correctly, since if she had only 9 questions correct, her score would have been 45, and no amount of incorrect answers could reduce her score to 44. If we begin with a guess of 10 correct answers, we can quickly see that if she had 3 wrong, she would reach a score of 44 as follows: 10 correct = +50, and 3 wrong = −6 for a total score of 44. Carol responded to 13 questions, and there were 7 questions left unanswered. But is there another answer? What if she had 11 questions correct? This would be impossible, since $11 \cdot 5 = +55$, and she could not arrive at a score of 44 by subtracting a multiple of 2. Suppose she had 12 questions correct. Then she would have had 8 questions wrong, which would give her the required 44, since $12 \cdot 5 = +60$, $8 \cdot -2 = -16$, and $60 - 16 = 44$. This would mean she had not omitted any questions, which was not the case. Any larger number of correct answers would clearly take her over the limit. Thus, the only possible number of omitted questions is 7. The intelligent guessing and testing technique reveals the answer quickly.

Logic Novelty 34

Problem: Four fair coins are tossed. What is the probability that at least two heads are face up?

Solution: In order to solve this problem, you may want to resort to one or more of the common permutation, combination, and probability formulas. However, the problem can be easily solved by making an exhaustive list showing all the possible cases, as seen in Figure 2-15, where H is a head and T is a tail. Notice that each column has the same number of tails, which enables us to keep "order."

H H H H	H H T T	H T T T	T T T T
H H H T	H T H T	T H T T	
H H T H	T H H T	T T H T	
H T H H	T H T H	T T T H	
T H H H	T T H H		
	H T T H		

Figure 2-15

There are a total of 16 possible cases. Of these, we can easily count the number of cases (11) with two or more heads. Therefore, the probability of getting at least two heads face up is $\frac{11}{16}$.

Logic Novelty 35

Problem: Christine has three dogs at home. When asked their ages, she replied that the product of their ages was 36. When pressed further, she added that "The sum of their ages is the same as the age of my son." Her friend stated that she knows the age of Christine's son, but still couldn't tell the ages of the three dogs. "I forgot to tell you," said Christine, "the youngest dog is a collie." How old is the youngest dog?

Solution: If you attempt this problem by setting up a system of equations, you will obtain the following equations: $xyz = 36$, and the son's age is $x + y + z = $ son's age.

Notice that we now have more variables than we do equations. However, we can still apply the technique of preparing an organized, exhaustive list. Begin by listing all the number triples whose product is 36. Notice that the list must be constructed in a systematic, organized way, so as not to miss any of the possibilities:

$$1 \cdot 1 \cdot 36 = 36$$
$$1 \cdot 2 \cdot 18 = 36$$
$$1 \cdot 3 \cdot 12 = 36$$
$$1 \cdot 4 \cdot 9 = 36$$
$$1 \cdot 6 \cdot 6 = 36$$
$$2 \cdot 2 \cdot 9 = 36$$
$$2 \cdot 3 \cdot 6 = 36$$
$$3 \cdot 3 \cdot 4 = 36.$$

These eight entries form the complete list of three integers whose product is equal to 36. One of these triples is our answer. Now let's move on to the second fact, that is, the *sum* of their ages, which are the following:

$$1 + 1 + 36 = 38$$
$$1 + 2 + 18 = 21$$
$$1 + 3 + 12 = 16$$
$$1 + 4 + 9 = 14$$
$$1 + 6 + 6 = 13$$
$$2 + 2 + 9 = 13$$
$$2 + 3 + 6 = 11$$
$$3 + 3 + 4 = 10.$$

If Christine's friend knew the son's age and still couldn't tell the ages of the dogs, what caused the problem? After all, if the son's age is 21, the dogs' ages would be 1, 2, and 18. If the son's age is 14, the dogs' ages would be 1, 4, and 9. Aha—the son's age must be 13, since this is the only age that could cause a problem (since there are two sets whose sum is 13). Since the *youngest* dog is a collie, the ages must be 6, 6, and 1, the only trio whose sum is 13 and which contains a "youngest" dog. The problem is now solved as the youngest dog is 1 year old.

Logic Novelty 36

Problem: A train traveling at the uniform rate of 30 mph enters a tunnel twice the length of the train. Six minutes elapse from the moment the first car enters the tunnel to the moment the last car clears the tunnel. What is the length of the train in miles?

Solution: From the moment the first car (the engine) of the train enters the tunnel, until the last car leaves the tunnel, six minutes (or $\frac{1}{10}$ hour) have elapsed. If the length of the train is x miles, then the length of the tunnel is $2x$ miles. Thus, the total distance the train travels is $3x = (30)\left(\frac{1}{10}\right) = 3$, so that $x = 1$ mile.

Logic Novelty 37

Problem: In an election, 5,219 votes were cast for four candidates. The winner exceeded his three opponents by 22 votes, by 30 votes, and by 73 votes, respectively. How many votes did the winner receive?

Solution: Using logic, one realizes that, of the 5,219 total votes, there were $22 + 30 + 73 = 125$ votes by which the winner exceeded the total number of votes that the three opponents received. Therefore, had all contestants received equal numbers of votes, that total would be $5,219 + 125 = 5,344$, with each of the four contestants receiving $\frac{5344}{4} = 1,336$ votes, which tells us that the winner must have received 1,336 votes.

Logic Novelty 38

Problem: If a clock indicates 6 o'clock by striking 6 times in 5 seconds, how many seconds does it take the clock to strike 12 times at noon? (Assume the striking consumes no time.)

Solution: If the first sound occurs at 6 o'clock, then an additional sound occurs for each of the next 5 seconds. Hence, there is a 1-second interval between sounds. At 12 o'clock, when the clock sounds 12 times, there are 11 intervals each requiring 1 second. Therefore, the clock strikes for 11 seconds at noon.

Logic Novelty 39

Problem: When dishtowels are hung to dry on a straight clothesline, each towel must have a clothespin at each end. However, the same pin can hold two towels when they hang next to each other. How many clothespins are needed to hang up a dozen towels?

Solution: The towels can have 11 overlapping edges, each requiring one pin. Each of the two free ends also needs a pin. Hence, 13 pins are needed.

Logic Novelty 40

Problem: If 9 carpenters can build a house in 36 days, how many days will it take 27 carpenters to build the same house?

Solution: To build the house, (9)(36) "labor-days" are required. Let x = the number of days in which 27 carpenters can build the same house. $27x = (9)(36)$, and $x = 12$.

Logic Novelty 41

Problem: Under what conditions can the following statement be true: $(x+y)^2 = x^2 + y^2$?

Solution: Just to begin with a simple replacement, suppose that $x = 1$ and $y = 1$, then the equality above will not hold since $(1 + 1)^2 \neq 1^2 + 1^2$. Therefore, the only way that the above equality can hold is if $x = 0$ and $y = 0$.

This can also be seen from an algebraic point of view since $(x + y)^2 = x^2 + y^2 + 2xy \neq x^2 + y^2$, and only can be equal when $x = 0$ and $y = 0$.

Logic Novelty 42

Problem: During which two-minute interval after 5 o'clock will the minute hand pass over the hour hand?

Solution: Assume that the hour hand will not move. In this case, the passing over will occur at 5:25. However, in actuality, the hour hand will move, so to compensate for this movement we multiply 25 by the "magic fraction"[2] $\frac{12}{11}$ to get $\frac{300}{11} = 27\frac{3}{11}$. Thus, the passing over occurs at $5:27\frac{3}{11}$.

You might also consider an alternate solution: At 5:24 the hour hand will be at the 27-minute mark, since it started from the 25-minute mark and traveled one mark every $\frac{1}{5}$ hour (12 minutes). For the next 12 minutes, the hour hand will be in the interval between 5:27 and 5:28. However, the minute hand will pass through this interval, in this time, thus passing over the hour hand.

Logic Novelty 43

Problem: A hungry hunter meets two farmers who have 8 loaves of bread. One farmer has 3 loaves and the other has 5 loaves. The farmers and the hunter agreed to divide the 8 loaves equally among the 3 participants, for which the hunter pays $8 for his share. What is the proper way for the farmers to divide the $8?

---

[2] See Arithmetic Novelty 36.

Solution: We consider the 8 loaves as part of the farmers' pot. (See Figure 2-16.)

	First farmer	Second farmer	Hunter
Contribution to the "pot"	3 loaves	5 loaves	0 loaves
Share taken from the "pot"	$\frac{8}{3}$ loaf	$\frac{8}{3}$ loaf	$\frac{8}{3}$ loaf
Amount left in the "pot"	$\frac{1}{3}$ loaf	$\frac{7}{3}$ loaf	

Figure 2-16

Since the hunter pays $8 for $\frac{8}{3}$ loaves, each loaf is worth

$$\frac{8}{\frac{8}{3}} - \frac{8}{\frac{8}{3}} \cdot \frac{3}{3} = 3 \text{ or } \$3.$$

The first farmer receives $\frac{1}{3}$ of $3, which is the value of the loaf. That is, he receives $1 for his contribution. The second farmer receives $\frac{7}{3}$ of the value of a loaf, that is, $\frac{7}{3}$ of $3 = $7 for his donation.

Logic Novelty 44

Problem: At dawn on the morning of July 1, Herbert is adrift on a raft 22 miles from shore. He can paddle 7 miles each day, but drifts back 2 miles each night while asleep. On what date will Herbert finally reach shore?

Solution: List the facts and restate the question. Herbert has 22 miles to go. He can advance 7 miles each day but will lose 2 miles each night. How many days will it take for Herbert to advance 22 miles? Once Herbert gets within 7 miles of land, he can reach the shore in a day without losing 2 miles at night. So, put aside 7 of the 22 miles for the final day's paddle. This leaves 15 miles to consider. Herbert's net gain each day is 5 miles (forward 7 miles and backward 2 miles). So, he can cover 15 miles in exactly 3 days. Therefore, he can reach shore in 3 + 1, or 4 days. This would be in the evening of July 4.

As a check, run through the scenario.

- Wakes up July 1st 22 miles from shore. Advances 7 miles; loses 2 miles.
- Wakes up July 2nd 17 miles from shore. Advances 7 miles; loses 2 miles.
- Wakes up July 3rd 12 miles from shore. Advances 7 miles; loses 2 miles.
- Wakes up July 4th 7 miles from shore. Advances 7 and reaches shore!

Logic Novelty 45

Problem: Together, Max, Sam, and Jack can do a piece of work in 8 hours. Max and Sam can do it in 12 hours, and Max does as much work as Sam and Jack together. How much time does each require to do the entire job alone?

Solution: If x, y, and z represent the times that Max, Sam, and Jack, respectively, can each complete the job working alone, then

- $\dfrac{1}{x}$ represents the part of the work completed by Max in 1 hour;
- $\dfrac{1}{y}$ represents the part of the work completed by Sam in 1 hour;
- $\dfrac{1}{z}$ represents the part of the work completed by Jack in 1 hour.

Therefore:

$$\frac{1}{x} + \frac{1}{y} + \frac{1}{z} = \frac{1}{8} \tag{a}$$

$$\frac{1}{x} + \frac{1}{y} = \frac{1}{12} \tag{b}$$

$$\frac{1}{x} = \frac{1}{y} + \frac{1}{z}. \tag{c}$$

Substitute (c) into (a): $\frac{1}{x} + \frac{1}{x} = \frac{1}{8}$; $x = 16$ (hours). Therefore, Max can do the entire job in 16 hours.

From (b) we get $\frac{1}{x} + \frac{1}{y} = \frac{1}{12}$, so that $\frac{1}{16} + \frac{1}{y} = \frac{1}{12}$ and then $y = 48$. So Sam can do the entire job in 48 hours.

From (c) we get $\frac{1}{x} = \frac{1}{y} + \frac{1}{z}$, which then is $\frac{1}{16} = \frac{1}{48} + \frac{1}{z}$, or $z = 24$. Finally, Jack can do the entire job in 24 hours.

Logic Novelty 46

Problem: You have a nickel, a dime, a quarter, and a fifty-cent piece. A clerk shows you several articles, each of which has a different price, any one of which you could buy with coins without receiving change. What is the largest number of articles the clerk could have shown you?

Solution: There are various ways to solve this problem. For a practical approach, consider the following:

- Using only one coin yields **5, 10, 25, and 50**.
- Using only two coins yields **5 + 10 = 15, 5 + 25 = 30, 5 + 50 = 55, 10 + 25 = 35, 10 + 50 = 60, and 25 + 50 = 75**.
- Using only three coins yields **5 + 10 + 25 = 40, 10 + 25 + 50 = 85, 5 + 25 + 50 = 80, and 5 + 10 + 50 = 65**.
- Using all four coins yields **5 + 10 + 25 + 50 = 90**.
- The items can cost any of the following: 5, 10, 15, 25, 30, 35, 40, 50, 55, 60, 65, 75, 80, 85, 90.

Since all the amounts are different from another, we can also consider the number of combinations. We can either use one of the coins or not, which means there are two possibilities for each coin. Therefore, there are $2 \cdot 2 \cdot 2 \cdot 2 = 16$ possibilities, one of which is that no coins were used, which must be eliminated. Thus, there are $16 - 1 = 15$ possible articles to purchase.

Logic Novelty 47

Problem: The size of a new camera is 25% smaller than the size of the old one, and it weighs $\frac{1}{2}$ as much. The weight per cubic unit of the new camera is how many times that of the old camera?

Solution: The new camera size is $\frac{3}{4}$ of the old camera size. The new camera weight is $\frac{1}{2}$ of the old camera weight. Hence, the new $\frac{\text{camera weight}}{\text{camera size}} = \left(\frac{\frac{1}{2}}{\frac{3}{4}}\right)$ of the old camera size is $\frac{2}{3}$ of the old camera size.

Logic Novelty 48

Problem: Evelyn is 3 years younger than Henry. The sum of their ages is greater than 30 and less than 35. What can Henry's age be?

Solution: There are two possibilities to consider here. To begin, we will let H = Henry's age.

Then $H - 3$ = Evelyn's age, so that the sum of their ages is $2H - 3$. Let us consider each of the two conditions separately: First, when the sum of their ages is greater than 30, we have $2H - 3 > 30$, so that $2H > 33$, and then $H > 16\frac{1}{2}$. This would indicate that Henry must be age 17 or older.

We now consider the second condition, where the sum of their ages is less than 35, which can be expressed as $2H - 3 < 35$, so that $2H < 38$, which gives us $H < 19$. Therefore, Henry must be age 18 or younger. By combining both conditions, we find that Henry must be either 17 or 18.

Logic Novelty 49

Problem: If a brick weighs 5 pounds and a half of a brick, then what is the weight of the brick expressed in pounds?

Solution: The wording of this problem could be somewhat confusing. The reader might interpret the weight of the brick to be 5 pounds plus half of the 5 pounds, in other words, $5 + 2.5 = 7.5$. This is not correct! Simple algebra can help us here, if we let w = the weight of the brick. Then, $w = 5 + \frac{1}{2}w$, so that $w = 10$.

Logic Novelty 50

Problem: We present a game that depends an understanding of prime numbers (numbers whose only factors are 1 and the number itself). The game has a clear strategy that will allow one player to win each time. It works as follows: Jack and Charlie have a pile of 150 toothpicks. The rules are that they take turns and must remove at least one or a prime number of picks each go. The person to take the last toothpick wins the game. Assuming Jack goes first, what is the strategy to win the game?

Solution: Jack would win the game by taking 2 toothpicks from the pile of 150, which leaves 148, and by taking 1, 2, or 3 toothpicks on each future turn. By doing this, Jack always leaves a number of toothpicks in the pile that is a multiple of 4 for Charlie to select from. Charlie can never take a multiple of 4 himself to switch tactics on Jack since it is not a prime number. By this strategy, Jack maintains control of the game and will eventually reach 0 and win.

Logic Novelty 51

Problem: If, on average, a hen and a half can lay an egg and a half in a day and a half, how many eggs should 6 hens lay in 8 days?

Solution: Since $\frac{3}{2}$ hens work for $\frac{3}{2}$ days, we may speak of the job as $\left[\frac{3}{2}\right]\left[\frac{3}{2}\right]$ or $\frac{9}{4}$ "hen-days." Similarly, the second job is $(6)(8)$ or 48 "hen-days." Let $x =$ the number of eggs laid by 6 hens in 8 days. Thus, we can form the following proportion:

$$\frac{\frac{9}{4}\text{ hen-days}}{48\text{ hen-days}} = \frac{\frac{3}{2}\text{ eggs}}{x\text{ eggs}}.$$

Cross multiply:

$$\frac{9}{4}x = (48)\left(\frac{3}{2}\right) = 72; \quad x = 32.$$

An alternate solution would be to set up the following structure:

$$\frac{3}{2} \text{ hens lay } \frac{3}{2} \text{ eggs in } \frac{3}{2} \text{ days.}$$

Then by doubling the number of hens and by keeping the number of days constant:

$$3 \text{ hens lay } 3 \text{ eggs in } \frac{3}{2} \text{ days.}$$

Now by doubling the number of days and by keeping the number of hens constant:

3 hens lay 6 eggs in 3 days.

Doubling the number of hens and keeping the number of days constant:

6 hens lay 12 eggs in 3 days.

In one day these 6 hens should produce 12 eggs:

6 hens lay 4 eggs in 1 day.

Therefore, in 8 days these 6 hens will produce $4 \cdot 8 = 32$ eggs.

6 hens lay 32 eggs in 8 days.

Logic Novelty 52

Problem: Find the next three terms in the following sequence: 0, 1, 2, 3, 6, 7, 14, 15, 30, __, __, __.

Solution: Disappointingly, a quick inspection shows no common difference between terms. However, imbedded in the sequence are several pairs of consecutive numbers: 0, 1; 1, 2; 2, 3; 6, 7; 14, 15, which should begin to give a clue about the pattern working backwards. One notices each pair, such as 14 and 15, begins with the number which is twice its predecessor. For instance, consider the pair of numbers beginning with 2, which is twice 1, and the next pair we find that the first member of the pair, 6, is twice 3, and the next pair's initial number 14 is twice 7, and 30 is twice 15. Thus, the next number after 30 would be its successor, 31, which we double to get 62 and then add 1 to get 63. Thus, the next three terms of the given sequence are 31, 62, 63.

Logic Novelty 53

Problem: A sequence of numbers can appear confusing when its pattern does not steadily increase or decrease, such as in the following problem. One term does not belong to the following sequence: 1, 8, 3, 6, 9, 4, 7, 2. Which number should be replaced, and by what number, in order to make the sequence consistent?

Solution: The up and down fluctuation of this sequence *could* signal two sequences interspersed. Hence, we will look at every other number to see if there is a pattern. The even-position terms build a sequence: 8, 6, 4, 2. This seems to be a complete and correct sequence. An inspection of the odd positioned terms shows the sequence 1, 3, 9, 7. If number 9 were replaced by 5, we would have consecutive odd numbers.

Logic Novelty 54

Problem: Ms. Williamson is forming committees in her class of 30 children. The committees may be formed with 4, 5, 7, or 8 children. What is the minimum number of committees she can form so that every child is on exactly one committee?

Solution: Since we are interested in minimizing the number of committees that Ms. Williamson wishes to form, she must try to form as many committees of 8 as possible. Suppose she creates three committees of 8. This accounts for 24 children. The remaining 6 children cannot fit in her committee scheme. She now tries using two committees of 8, leaving 14 children to be placed on committees. The least number of committees for these 14 would be two committees of 7 each. Therefore, the minimum number of committees would be 4.

Logic Novelty 55

Problem: A hotel issues parking permits consisting of 2 letters (not including O) followed by 3 numerals with no numeral repeated. How many different permits can they print?

Solution: We will look at the number of permits by systematically counting the number of possible permit numbers. Consider the 5 positions to be filled. The first 2 can each be filled in any one of 25 different ways, since there are 25 letters to choose from (remember that the letter O is excluded). The next position can be filled in 10 different ways since there are 10 digits to choose from. With no digit permitted to be repeated, the fourth place can be filled in any one of 9 ways, and the fifth position can be filled in any one of 8 ways. The product of these positions is $25 \times 25 \times 10 \times 9 \times 8 = 450{,}000$, which is the number of permits which can be printed with these restrictions.

Logic Novelty 56

Problem: In a room with six people, each person shakes hands with another person exactly once. How many handshakes are there?

Solution: One may draw a diagram representing six people and draw lines to indicate handshakes. This could lead to a successful counting procedure if done systematically. However, this particular problem can be solved by organized counting in the following way. The first person shakes hands with 5 others. The second person, having shaken the first person's hand, shakes hands with 4 other people. The third person shakes hands with 3 other people. The fourth person with 2 others, and the fifth person with 1 other person, making a total of $5 + 4 + 3 + 2 + 1 = 15$ handshakes. Note that the sixth person is not considered since he already has shaken everybody else's hand.

Logic Novelty 57

Problem: Four college roommates, Al, Ben, Charlie, and David, baked cookies to bring home when they left for holiday break the next day. The next morning, Al woke up first. He took $\frac{1}{4}$ of the cookies and left for home. Ben woke up next and took $\frac{1}{4}$ of the cookies that remained and left. Next, Charlie took $\frac{1}{4}$ of the remaining cookies and left. Finally, David took $\frac{1}{4}$ of the remaining cookies. There were 81 cookies left after David took his. How many cookies had the roommates baked?

Solution: List the facts and restate the question. Four roommates baked cookies. One at a time, each took $\frac{1}{4}$ of the cookies that remained. After everyone had taken cookies, 81 cookies remained. How many cookies had the roommates baked?

There were 81 cookies left after David took $\frac{1}{4}$. So, 81 is $\frac{3}{4}$ of the number of cookies David found when he woke up. Because $\frac{3}{4} \cdot 108 = 81$, there must have been 108 cookies when David got to them. There were 108 cookies left after Charlie took $\frac{1}{4}$. Because $\frac{3}{4} \cdot 144 = 108$, there must have been 144 cookies when Charlie got to them. There were 144 cookies left after Ben took $\frac{1}{4}$. Because $\frac{3}{4} \cdot 192 = 144$, there must have been 192 cookies when Ben got to them. There were 192 cookies left after Al took $\frac{1}{4}$. Because $\frac{3}{4} \cdot 256 = 144$, there must have been 256 cookies when Al got to them. So, the roommates baked 256 cookies.

Check to make sure your answer makes sense as follows: Assume the roommates baked 256 cookies and play out the scenario.

- Al took $\frac{1}{4}$ of 256 cookies, or 64 cookies. So, $256 - 64 = 192$ cookies remained.
- Ben took $\frac{1}{4}$ of 192 cookies, or 48 cookies. So, $192 - 48 = 144$ cookies remained.
- Charlie took $\frac{1}{4}$ of 144 cookies, or 36 cookies. So, $144 - 36 = 108$ cookies remained.
- David took $\frac{1}{4}$ of 108 cookies, or 27 cookies. So, $108 - 27 = 81$ cookies remained.

The problem states that 81 cookies remained after David took some, so our answer is correct.

Logic Novelty 58

Problem: A circular portion of the schoolyard has been set aside for the different grades to plant their gardens. The circle is divided up into different size regions by stretching ropes across the circle. What is the maximum number of regions that can be obtained by using 7 ropes?

Solution: One could begin attempting to solve this problem by drawing a picture of the situation by progressively increasing the number of ropes separating the schoolyard. In Figure 2-17, we show a progressive increase till four ropes are used to divide the schoolyard. Just drawing four ropes is already something of a challenge, so one can imagine drawing seven ropes would be rather complicated.

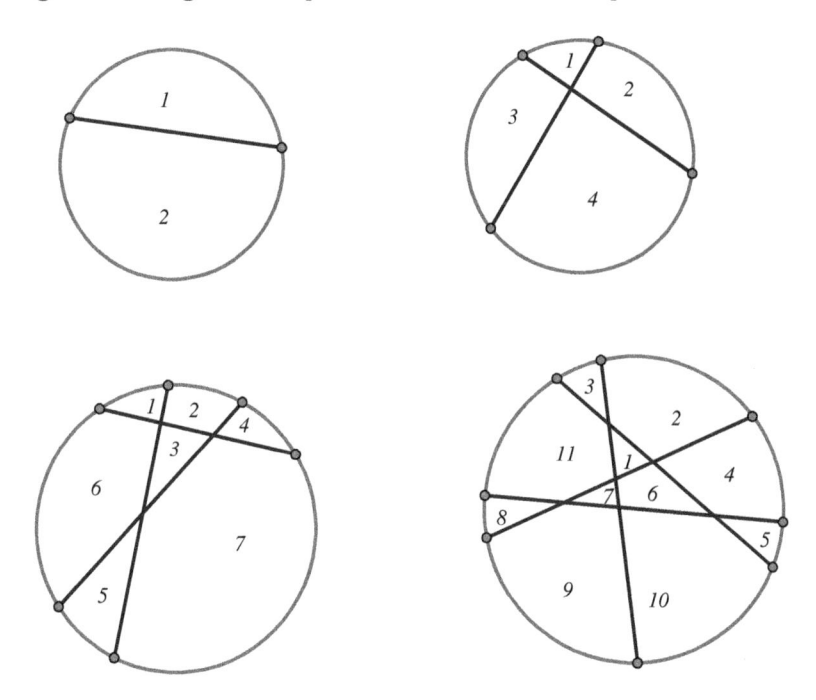

Figure 2-17

A sophisticated mathematics enthusiast might recognize that there could be a formula that could generate the desired result. The formula for the number of regions is $y = \frac{x^2 + x + 2}{2}$, where y = the maximum number of regions obtained with x ropes. Of course, it is rather unlikely that you would be familiar with this formula, so we need another method. Let's reduce the problem to simpler cases, keeping track as we increase them, and look for a pattern to emerge. We can use the table in Figure 2-18 to observe our results as the number of ropes increases.

Number of ropes	0	1	2	3	4	5	6	7
Maximum number of regions	1	2	4	7	11	16	22	29
Difference		1	2	3	4	5	6	7

Figure 2-18

The differences between numbers of regions increases by one each time. Now we can find the number of regions generated when seven ropes are used to partition the schoolyard: 29. Perhaps you can imagine how difficult it would be to use the diagram to discover this result. The use of patterns is critical strategy in problem solving. In fact, pattern recognition is often combined with other strategies in order to solve problematic situations.

Logic Novelty 59

Problem: Rosie has an 11-liter can and a 5-liter can. How can she measure out exactly 7 liters of water?

Solution: One might simply guess at the answer, and keep pouring water back and forth between the two cans in an attempt to arrive at the correct answer, unsystematically guessing and testing. However, the problem can be solved in a more organized manner by using a working-backwards strategy. We need to end up with 7 liters in the 11-liter can, leaving a total of 4 empty liters in the can. But where do 4 empty liters come from? (See Figure 2-19.)

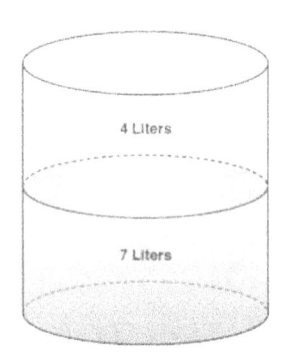

4 Liters

7 Liters

Figure 2-19

To obtain 4 liters, we must leave 1 liter in the 5-liter can. Now, how can we obtain 1 liter in the 5-liter can? Fill the 11-liter can and pour from it twice into the 5-liter can, discarding the water. This leaves 1 liter in the 11-liter can. Pour the 1 liter into the 5-liter can. (See Figure 2-20.)

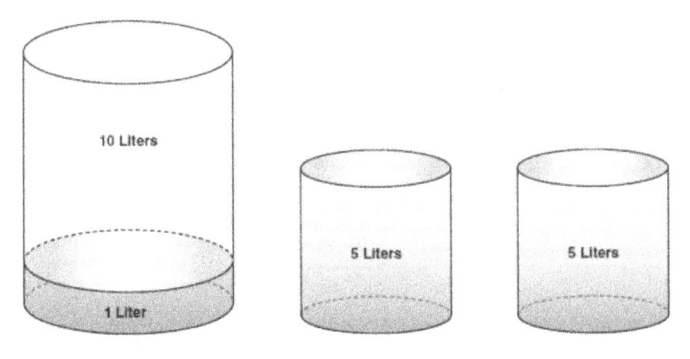

Figure 2-20

Now, fill the 11-liter can and pour off the 4 liters needed to fill the 5-liter can. This leaves the required 7 liters in the 11-liter can. (See Figure 2-21.)

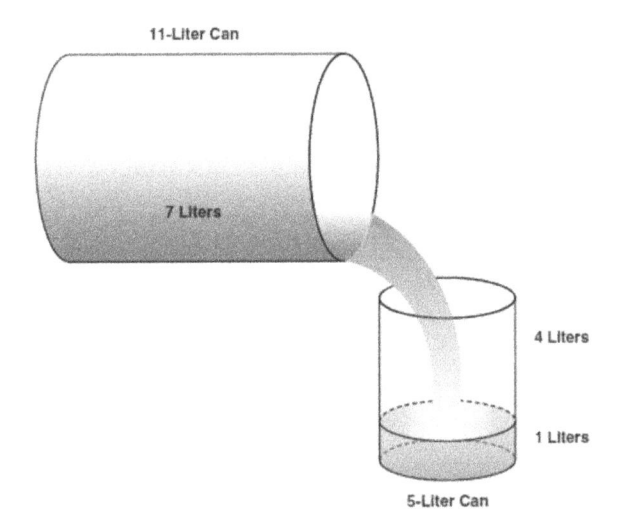

Figure 2-21

Note that problems of this sort do not always have a solution. That is, if you wish to construct additional problems of this sort, you must bear in mind that a solution only exists when the difference of multiples of the capacities of the two given cans can be made equal to the desired quantity. In this problem, $2(11) - 3(5) = 7$.

Logic Novelty 60

Problem: One member of the RW Society, which has 100 members, has been notified that the Society's meeting place must be changed. This member activates the Society's telephone squad by telephoning three other members, each of whom then telephones another three members and so on, until all 100 RW Society members have been notified of the meeting place change. What is the greatest number of members of the RW Society who do not need to make a telephone call?

Solution: The typical response to this question is to simulate the action in the problem by counting. We will begin to count systematically:

- Member #1 calls three other members; total members contacted = 4.
- Members #2, #3, and #4 each make 3 calls; total members contacted = $4 + 9 = 13$.
- Members #5 – #13 each make 3 calls; total members contacted = $4 + 9 + 27 = 40$.
- Members #14 – #33 each make 3 calls; total members contacted = $4 + 9 + 27 + 60 = 100$.

Since 33 members had to make telephone calls to reach all 100 members, there were 67 members who did not have to make any calls.

This problem can be solved in a simpler way by *working backwards.* We know that, after the first member has been contacted, there are 99 additional members who have to be contacted. This requires 33 members each making 3 telephone calls. This leaves 67 members who need not make any calls.

Logic Novelty 61

Problem: Jimmy, Ronald, George, and Bill spent the afternoon picking apples, which they put into a large box. They were so tired from their labors that they went to sleep early that evening. During the night, Jimmy awoke, took $\frac{1}{4}$ of the apples from the box, put them in his backpack, and went back to sleep. Later, Ronald awoke, took $\frac{1}{4}$ of the apples from the box, put them into his backpack, and went back to sleep. Later, George awoke, took $\frac{1}{4}$ of the apples from the box and put them into his backpack. He, too, went back to sleep. Finally, just before daylight, Bill awoke, took $\frac{1}{4}$ of the apples from the box, put them into his backpack and woke the other three boys. They counted the apples and found 81 apples in the box. How many apples had they picked?

Solution: We can write an equation to find the number of apples as follows. We let x represent the number of apples originally picked and placed in the box.

- Jimmy took $\frac{1}{4}$ of x, leaving $\frac{3}{4}$ of x in the box.
- Ronald then took $\frac{1}{4}$ the $\frac{3}{4}$ of x that remained, leaving $\frac{3}{4}$ of $\frac{3}{4}$ of x.
- Then George took $\frac{1}{4}$ of the $\frac{3}{4}$ of the $\frac{3}{4}$ of x that remained.
- Finally, Bill took $\frac{1}{4}$ of the $\frac{3}{4}$ of the $\frac{3}{4}$ of the $\frac{3}{4}$ of x that remained.

Thus, $\frac{3}{4} \cdot \frac{3}{4} \cdot \frac{3}{4} \cdot \frac{3}{4} x = \frac{81}{256} x = 81$, so that $x = 256$, therefore, they picked 256 apples.

Although this procedure seems satisfactory, finding the correct equation can prove somewhat difficult. Let's use a working-backwards strategy. Bill left 81 apples after taking $\frac{1}{4}$. This was $\frac{3}{4}$ of the apple sum, therefore $\frac{4}{4} = 108$. George left 108 apples after taking $\frac{1}{4}$. This was $\frac{3}{4}$ of the apple sum, therefore, $\frac{4}{4} = 144$. Ronald left 144 apples after taking $\frac{1}{4}$. This was $\frac{3}{4}$ of the apple sum, therefore, $\frac{4}{4} = 192$. Jimmy left 192 apples after taking $\frac{1}{4}$. This was $\frac{3}{4}$ of the apple sum, therefore, $\frac{4}{4} = 256$. Thus, they began with 256 apples.

Logic Novelty 62

Problem: The month of February had 5 Sundays in 2004. In which other years during the 21st century will the month of February again have five Sundays?

Solution: February can only have five Sundays if February 1st falls on a Sunday *and* if the year is a leap year. We use a strategy of logical reasoning to determine that both of these conditions are met every 28 years because there are 7 different days in the week and every four years we have a leap year. The least common multiple of 7 and 4 is 28. Therefore, 2004 + 28 = 2032, and 2032 will be the next year with five Sundays in February. It follows that 2032 + 28 = 2060 will also be such a year, as well as 2060 + 28 = 2088.

Logic Novelty 63

Problem: Max has between 50 and 100 pennies. When he stacks them in piles of 2, he has one penny left over. When he stacks them in piles of 3, he has one penny left over. When he stacks them in piles of 4, he has one penny left over. However, when he stacks them in piles of 5, there are no pennies left over. How many pennies does Max have?

Solution: The strategy that we will use here is clearly mapped out for us. We will organize the data as was given:

1. When divided by 2 (or grouped by 2s), the remainder is 1.
2. When divided by 3, the remainder is 1.
3. When divided by 4, the remainder is 1.
4. When divided by 5, the remainder is 0.
5. There are between 50 and 100 coins (exclusively).

Clue 4 tells us that the number has to be a multiple of 5. Therefore, we consider all possible multiples of 5 between 50 and 100: 55, 60, 65, 70, 75, 80, 85, 90, 95.

Clue 1 tells us that the number is an odd number (because would it be even, then there would not be a remainder of 1 when divided by 2). So, we can eliminate all even numbers we considered for clue 4 and are left with: 55, 65, 75, 85, 95.

Clue 2 gives us the following information: The number has to be one more than a multiple of 3. This leaves only two choices: 55 or 85.

Clue 3 tells us that we can eliminate 55, since 55 divided by 4 does not leave a remainder of 1.

Therefore, the only number satisfying all clues is 85. So, Max has 85 pennies.

Logic Novelty 64

Problem: If 100 more than the sum of n consecutive integers is equal to the sum of the next n consecutive integers, find n.

Solution: Since each integer in the second sum is n more than the corresponding integer in the first sum, and since there are n integers in each sum, then the first sum +100 is equal to the second sum plus n^2.

Symbolically, this can be solved as:

$$\frac{n}{2}\left[a+(a+n-1)\right]+100=\frac{n}{2}\left[(a+n)+(a+n+n-1)\right],$$

which can be simplified to $n^2 = 100$, so that $n = 10$.

Logic Novelty 65

Problem: A grocer bought some lemons at the rate of 3 for 16¢. He bought twice as many lemons at the rate of 4 for 21¢. To make a profit of 20% based on his investment, he sold all of them at the rate of 3 for n cents.

Solution: This is best solved symbolically as follows:

$$\frac{\frac{16}{3}+2\cdot\frac{21}{4}}{3}\cdot\frac{6}{5}=120\% \text{ of the cost}=\frac{19}{3}.$$

This would indicate that the grocer should sell three lemons for 19 cents.

Logic Novelty 66

Problem: An integer is represented by a two-digit decimal. If three times the sum of the digits is added to the integer, its digits are reversed. Find all such positive integers.

Solution: This is best solved symbolically by having t represent the tens digit and having u represent the units digit. We then have the following:

$$3(t + u) + 10t + u = 10u + t, \quad \text{whereupon } 2t = u.$$

Therefore, the only two-digit numbers that satisfy this relationship are: 12, 24, 36, and 48.

Logic Novelty 67

Problem: After finding the average of 25 scores, a student carelessly included the average with the 25 scores and found the average of these 26 numbers. How does the average of the 26 numbers compare to the average of the original 25 numbers?

Solution: If we let A equal the average of the first 25 numbers and B equal the average of the second batch of 26 numbers, then $B = \frac{1}{26}(25A + A) = \frac{1}{26}(26A) = A$, which indicates that both averages were the same.

Logic Novelty 68

Problem: Two candles of the same length are made of different materials so that one burns out completely at a uniform rate in 3 hours and the other candle burns out completely in 4 hours. At what time in the afternoon should the candles be lit so that at 4 pm one stub is twice the length of the other?

Solution: Let the length of the candle be chosen as the unit of length. Let t represent the number of hours before 4 pm needed to produce the desired result. In one hour, the faster burning candle shortens by $\frac{1}{3}$, the slower candle shortens by $\frac{1}{4}$ is length, and in t hours, they shortened by $\frac{t}{3}$ and $\frac{t}{4}$, so their lengths are $1 - \frac{t}{3}$, and $1 - \frac{t}{4}$, respectively.

Then $\left(1-\frac{t}{4}\right)=2\left(1-\frac{t}{4}\right)$, so that $t=2\frac{2}{5}$ hours before 4 pm. The time for the candles to be lit is, therefore, $4-2\frac{2}{5}=1\frac{3}{5}$ hours after noon, or at 1:36 pm.

Logic Novelty 69

Problem: Find the greatest common denominator of 6,432 and 132 diminished by 8.

Solution: Consider the prime factors of each number, which are $6{,}432 = 2^5 \cdot 3 \cdot 67$, and $132 = 2^2 \cdot 3 \cdot 11$. We can see that the common factors are $2^2 \cdot 3 = 12$ and when we deduct 8 the result is 4.

Logic Novelty 70

Problem: Geometry problems can also be solved by using simple logic. Consider circle O where arc DE is 90°, as shown in Figure 2-22. The base BC of isosceles triangle ABC ($AB = AC$) contains the center of the circle O. What is the relationship between the area of triangle ABC in the area triangle DOE?

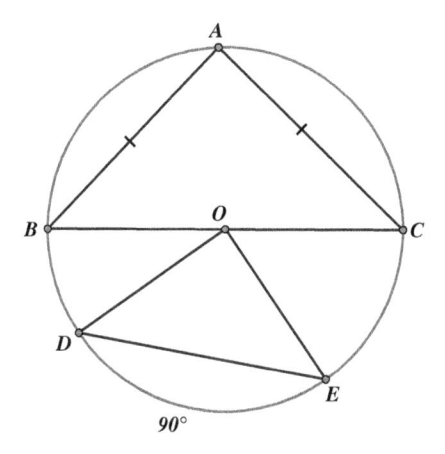

Figure 2-22

Solution: There are many ways to solve this problem. However, simple logic would have us move/rotate triangle *DOE* so that points *B* and *D* coincide. We show this in Figure 2-23. It should then be obvious that triangle *DOE* is half the size of triangle *ABC*. Simple logic has saved us significant geometric configurations.

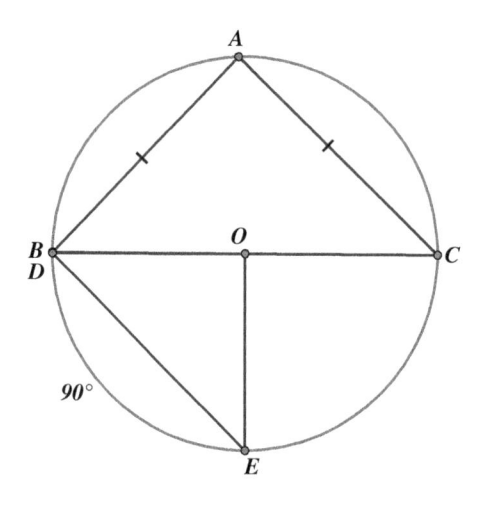

Figure 2-23

Logic Novelty 71

Problem: A man lives in the building located five blocks south and five blocks west of the building in which his sister lives. In driving to his sister's home, he drives 10 blocks. All the streets are in a rectangular pattern and all available for driving. In how many different ways can he drive from his home to his sister's home by driving only 10 blocks?

Solution: Let E denote the act of driving one block east and let N denote the act of driving one block north. Therefore, to each route there corresponds a string of 5 E's and 5 N's arranged in a row.

The number of ways one can fill 10 spaces with 5 E's and the remainder being 5N's is given by the number of combinations, which is $\frac{10!}{5! \cdot 5!} = 252$ ways.

Logic Novelty 72

Problem: A person has 2 pennies, 2 nickels, 2 dimes, and 2 quarters. In how many ways can the person select 2 of these coins whose value is an even number of cents?

Solution: Two coins can be chosen from the 8 available coins in $_8C_2 = \frac{8!}{6!2!} = 28$ ways. The only way to get an odd total is to choose a dime +1 coin other than a dime, since the dime is the only even number of cents. This can be done in $6 + 6$ ways. The number of ways to get an even total is therefore $28 - 12 = 16$.

Logic Novelty 73

Problem: The are at least two ways of doing this problem: the poet's way and the peasant's way. Which will you select? Here is the problem: What is the sum of all the numbers formed by rearranging the digits of the number 975?

Solution: The peasant's way to solve this problem is to list all the possible numbers that can be formed with the digits of the number 975. They are: 579, 597, 759, 795, 957, and 975. Then all one needs to do is to add the numbers and their sum is 4,662.

 The poet's curious approach to this problem is to note that each of the digits must appear twice in each of the place value positions, as can be seen above. Therefore, the sum of the digits in the sum must be $2(9 + 7 + 5) = 42$ in each place. We can, therefore, get the required sum by $42 \cdot 100 + 42 \cdot 10 + 42 = 4{,}662$.

Logic Novelty 74

Problem: The following problem requires some open thinking and logical reasoning.[3] In the configuration shown in Figure 2-24, there are 11 sticks in each outside row and column. How can you remove one stick from each outside row, and one from each outside column, and move the remaining sticks so as to end with 11 sticks in each of these rows and columns?

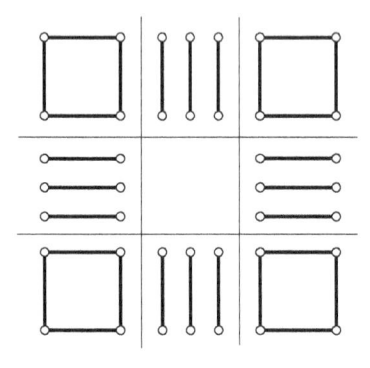

Figure 2-24

Solution: Many people become frustrated quickly as they attempt to solve this problem. In removing one stick from each row and column, they end up with the configuration shown in Figure 2-25.

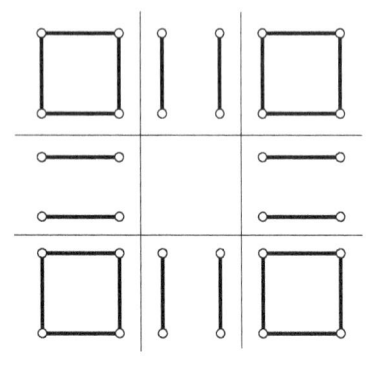

Figure 2-25

[3] More examples of this kind can be found in Alfred S. Posamentier and Ingmar Lehmann, *Magnificent Mistakes in Mathematics* (Amherst, New York: Prometheus Books, 2013), pp. 217–221.

However, in Figure 2-25 there remain only 10 sticks in each outside row and outside column. The question is how can we have 11 sticks in each outside row and outside column after we have removed one stick from each row and column? Once again, we need to think "out-of-the-box."

We simply take another stick from each row and column, as shown in Figure 2-26, and place one at each of the corners, and the problem is solved. This may appear trivial, but experience shows that most people do not "stumble" easily onto this curious solution.

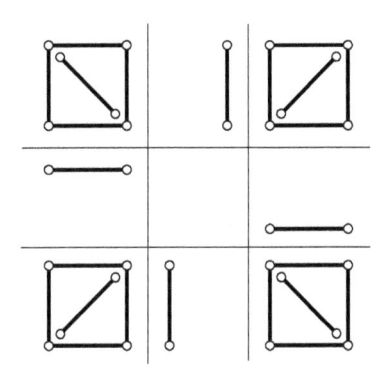

Figure 2-26

Logic Novelty 75

Problem: The following letters represent the digits of a simple addition:

$$
\begin{array}{r}
\text{SEND} \\
+\ \text{MORE} \\
\hline
\text{MONEY}
\end{array}
$$

Find the digits that represent the letters to make this addition correct. Show that the solution is unique.

Solution: The sum of two four-digit numbers cannot yield a number greater than 19,999. Therefore **M = 1**. We then have MORE < 2,000 and SEND < 10,000. It follows that MONEY < 12,000. Thus, O can be either 0 or 1. But the 1 is already used; therefore, **O = 0.**

$$\begin{array}{r} \text{S E N D} \\ .\quad \text{1 0 R E} \\ \hline \text{1 0 N E Y} \end{array}$$

Now MORE < 1,100. If SEND were less than 9,000, then MONEY < 10,100, which would imply that N = 0. But this cannot be since 0 was already used; therefore SEND > 9,000, so that **S = 9**. We now have:

$$\begin{array}{r} \text{9 E N D} \\ .\quad \text{1 0 R E} \\ \hline \text{1 0 N E Y} \end{array}$$

The remaining digits with which we may complete the problem are {2,3,4,5,6,7,8}.

Examining the units digits, we find that the greatest sum is $7 + 8 = 15$ and the least sum is $2 + 3 = 5$. If $D + E < 10$, then $D + E = Y$ with no carry over into the tens column. Otherwise, $D + E = Y + 10$, with a 1 carried over to the tens column. Taking this argument one step further to the tens column, we get $N + R = E$, with no carryover, or $N + R = E + 10$, with a carryover of 1 to the hundreds column. However, if there is no carryover to the hundreds column, then $E + 0 = N$, which implies that $E = N$. This is not permissible. Therefore, there must be a carryover to the hundreds column. So, $N + R = E + 10$, and $E + 0 + 1 = N$, or $E + 1 = N$. Substituting this value for N into the previous equation we get $(E + 1) + R = E + 10$, which implies that $R = 9$. But this has already been used for the value of S. We must try a different approach. We shall assume, therefore, that $D + E = Y + 10$, since we apparently need a carryover into the tens column, where we just reached a dead end.

Now the sum in the tens column is $1 + 2 + 3 < 1 + N + R < 1 + 7 + 8$. If, however, $1 + N + R < 10$, there will be no carryover to the hundreds column, leaving the previous dilemma of $E = N$, which is not allowed. We then have $1 + N + R = E + 10$, which insures the needed carryover to the hundreds column. Therefore, $1 + E + 0 = N$, or $E + 1 = N$.

Substituting this in the above equation $(1 + N + R = E + 10)$ gives us $1 + (E + 1) + R = E + 10$, or **R = 8**.

We now have:

$$
\begin{array}{r}
9END \\
.\ \ \ 108E \\
\hline
10NEY
\end{array}
$$

From the remaining list of available digits, we find that $D + E < 14$. So from the equation $D + E = Y + 10$, Y is either 2 or 3. If $Y = 3$, then $D + E = 13$, implying that the digits D and E can take on only 6 or 7. If $D = 6$ and $E = 7$, then from the previous equation $E + 1 = N$, we would have $N = 8$, which is unacceptable since $R = 8$. If $D = 7$ and $E = 8$, then from the previous equation $E + 1 = N$, we would have $N = 9$, which is unacceptable since $S = 9$. Therefore, **Y = 2**. We now have:

$$
\begin{array}{r}
9END \\
.\ \ \ 108E \\
\hline
10NE2
\end{array}
$$

Thus, $D + E = 12$. The only way to get this sum is with 5 and 7. If $E = 7$, we once again get from $E + 1 = N$, the contradictory $N = 8$, which is not acceptable. Therefore, **D = 7** and **E = 5**. We can now again use the equation $E + 1 = N$ to get **N = 6**. Finally, we get the solution:

$$
\begin{array}{r}
9567 \\
.\ \ \ 1085 \\
\hline
10652
\end{array}
$$

This rather strenuous activity should provide some important training in strengthening one's facility in logical thinking in mathematics.

Chapter 3

Algebraic Novelties

In school, algebra is introduced as the basic tool for understanding most aspects of mathematics. However, there is plenty of intrigue and beauty within the field of algebra that is usually omitted in the basic school curriculum. In this section, we will explore applications of algebra that will hopefully enhance a reader's appreciation of this most essential aspect of mathematics. Of course, we will use this important mathematical tool to solve various conundrums and observe how facility with algebra helps explain initially confusing issues. Essentially, this section will present challenging problems that can be nicely solved algebraically.

Algebraic Novelty 1

Problem: A certain fish has a head 7 inches long. Its tail is as long as its head and one-half as long as his body. The length of its body equals the length of its head and tail together. Our challenge is to find the length of the entire fish using simple algebra.

Solution: Let t = the length of the fish's tail and let b = the length of the fish's body; we have the head length as $h = 7$. We then have $t = 7 + \frac{1}{2}b$, and $b = 7 + t$. We then substitute for b in the first equation to get: $t = 7 + 2(7 + t)$, and then $t = 21$. Therefore, $b = 7 + 21 = 28$. Hence, the length of the entire fish = $h + b + t = 7 + 28 + 21 = 56$.

Algebraic Novelty 2

Problem: Here is a computation that can be simplified through algebra. How many twelfths of one yard is 60% of one foot?

Solution: We begin by letting x equal the sought after twelfths of the yard, which for our computation we will refer to as 3 feet. Therefore, $\frac{x}{12}$ of 3 feet $= 60\%$ of 1 foot. This can be written as $\frac{3x}{12} = \frac{3}{5}(1)$. Solving this equation gives us $x = \frac{12}{5} = 2\frac{2}{5}$ twelfths of a yard.

Algebraic Novelty 3

Problem: In an election, 6,523 votes were cast for four candidates. The winner exceeded his 3 opponents by 22 votes, by 30 votes, and by 73 votes, respectively. How many votes did the winner receive?

Solution: Let W = the number of votes the winner received. Then his 3 runners-up received $(W - 22)$, $(W - 30)$, and $(W - 73)$ votes, respectively. Therefore, we have the total number votes represented as follows: $4W - 125 = 6,523$, so that $W = 1,662$. Another simplified solution with simple algebra.

Algebraic Novelty 4

Problem: The combined weight of a bottle and a tumbler equals the weight of a pitcher. The weight of the bottle equals the weight of a plate and a tumbler. Two pitchers weigh the same as three plates. How many tumblers will balance with one bottle?

Solution:
- Let b = the weight of the bottle.
- Let t = the weight of the tumbler.
- Let p = the weight of the pitcher.
- Let r = the weight of the plate.

Representing the above given information in symbolic form we get the following equations:

$$b + t = p \tag{1}$$

$$b = r + t, \text{ or } r = b - t \tag{2}$$

$$2p = 3r \tag{3}$$

We substitute equations (1) and (2) into equation (3) to get $2(b + t) = 3(b - t)$, whereupon we find that $b = 5t$. Therefore, 1 bottle will balance with 5 tumblers.

Algebraic Novelty 5

Problem: We are given that $c > a > 0$, and $a - b + c = 0$. Find the larger root of $ax^2 + bx + c = 0$.

Solution: We are given that $b = a + c$, which when substituted in the given equation yields

$$ax^2 + (a + c)x + c = (ax + c)(x + 1) = 0.$$

Whereupon the roots are

$$x = -\frac{c}{a}, \text{ and } x = -1.$$

Since $c > a > 0$, we have $x = -\frac{c}{a}$, which is less than -1. Therefore, the larger root is $x = -1$.

An alternate solution uses the general quadratic formula

$$x = \frac{-b \pm \sqrt{b^2 - 4ac}}{2a}$$

so that the two roots in this particular case are

$$x = \frac{-(a+c) \pm \sqrt{(a+c)^2 - 4ac}}{2a} = \frac{-(a+c) \pm \sqrt{(a-c)^2}}{2a} = \frac{-(a+c) \pm |a-c|}{2a}.$$

Since $c > a$, $|a - c| = c - a$, the larger root is -1.

Algebraic Novelty 6

Problem: Find $f(x) = ax^2 + bx + c$, if $f(x + 3) = x^2 + 7x + 4$.

Solution: We begin by letting $y = x + 3$, so that $x = y - 3$, whereupon $f(y) = (y - 3)^2 + 7(y - 3) + 4 = y^2 + y - 8$. Therefore, we have $f(x) = x^2 + x - 8$.

Algebraic Novelty 7

Problem: What is the largest possible value for the sum $x + y + z$, where the following holds true:

$$x^2 + xy + xz = 16$$

$$y^2 + yz + xy = 24$$

$$z^2 + xz + yz = 9.$$

Solution: A quick review of each of these equations would lead us to see how we can factor the three equations to reach the desired sum, namely, $x + y + z$, as follows:

$$x(x + y + x) = 16$$

$$y(x + y + x) = 24$$

$$z(x + y + x) = 9.$$

When we add these three equations, we get the following sum: $(x + y + x)(x + y + x) = (x + y + x)^2 = 49$, so that $x + y + z = \pm 7$, thus, the larger sum is +7.

Algebraic Novelty 8

Problem: The challenge here is to write the expression $\sqrt{5 + \sqrt{24}}$ in the form of $\sqrt{n} + \sqrt{m}$, and then to find n and m, which are integers with $n > m$.

Solution: We let $\sqrt{5 + \sqrt{24}} = \sqrt{n} + \sqrt{m}$ and then square both sides to get $5 + \sqrt{24} = n + m + 2\sqrt{nm}$. Equating rational and irrational parts gives us the equations $5 = n + m$, and $\sqrt{24} = 2\sqrt{nm}$, which can be further simplified by squaring both sides to get $24 = 4(nm)$, or $6 = nm$. It is then obvious that $n = 3$ and $m = 2$.

Algebraic Novelty 9

Problem: Find the smaller of two numbers which are real, equal, and nonnegative, whose square of their sum is equal to their sum of squares.

Solution: Consider the two numbers a and b so that the sum of squares $a^2 + b^2 = (a + b)^2$.

Since $(a + b)^2 = a^2 + b^2 + 2ab$, the previous relationship can now be written as $a^2 + b^2 = a^2 + b^2 + 2ab$, which can only be true if $2ab = 0$. Since neither a nor b is negative, 0 must be the smallest of the two numbers.

Algebraic Novelty 10

Problem: If $x - y$ is a factor of $x^3 + y^3 + k(x^2 + y^2)(x + y)$, find the numerical value of k.

Solution: Consider the equation $x^3 + y^3 + k(x^2 + y^2)(x + y) = 0$ and consider x as the variable and consider y as the constant. Then $x - y$ is a factor of the left side of the equation if, and only if, $x = y$ is a root of the equation. That is, $x^3 + x^3 + k(x^2 + x^2)(y + y) = 2y^3 + 4ky^3 = 0$, whereupon it is clear that $k = -\frac{1}{2}$.

Algebraic Novelty 11

Problem: When 6 people work 7 hours per day for 10 days they produce 102 articles. How many articles of the same type can be produced by 7 people working at the same rate for 8 hours per day for 5 days?

Solution: In the first case, we have $6 \cdot 7 \cdot 10 = 420$ "people-days" producing 102 articles. In the second case, we have $7 \cdot 8 \cdot 5 = 280$ people-days. This allows us to set up the following proportion $\frac{420}{280} = \frac{102}{x}$, or $x = 68$ articles.

Algebraic Novelty 12

Problem: Find the value of a, b, and c that satisfies the three following equations:

$$\frac{1}{a}+\frac{1}{b}=7, \quad \frac{1}{b}+\frac{1}{c}=11, \quad \frac{1}{a}+\frac{1}{c}=13.$$

Solution: By adding the three equations together we get

$$2\left(\frac{1}{a}+\frac{1}{b}+\frac{1}{c}\right)=31, \text{ or } \frac{1}{a}+\frac{1}{b}+\frac{1}{c}=\frac{31}{2}.$$

Now we use some clever algebraic techniques by subtracting each of the original equations separately from their sum:

$$\left(\frac{1}{a}+\frac{1}{b}+\frac{1}{c}\right)-\left(\frac{1}{a}+\frac{1}{b}\right)=\frac{1}{c}=\frac{31}{2}-\frac{14}{2}=\frac{17}{2},$$

so that $c=\dfrac{2}{17}$.

$$\left(\frac{1}{a}+\frac{1}{b}+\frac{1}{c}\right)-\left(\frac{1}{b}+\frac{1}{c}\right)=\frac{1}{a}=\frac{31}{2}-\frac{22}{2}=\frac{9}{2},$$

then $a=\dfrac{2}{9}$.

$$\left(\frac{1}{a}+\frac{1}{b}+\frac{1}{c}\right)-\left(\frac{1}{a}+\frac{1}{c}\right)=\frac{1}{b}=\frac{31}{2}-\frac{26}{2}=\frac{5}{2},$$

and finally we have $b=\dfrac{2}{5}$.

Algebraic Novelty 13

Problem: Express in terms of n the sum of all numbers in the form of $3k+3$, as k takes on all integral values from 1 to n inclusive.

Solution: We can write the given format as

$$[2(1)+3]+[2(2)+3]+[2(3)+3]+[2(4)+3]+\cdots+[2(2)+3]$$

$$=2(1+2+3+4+\cdots+n)+3n=2\left(\frac{n(n+1)}{2}\right)+3n=n^2+n+3n=n^2+4n.$$

Remember that the sum of the consecutive numbers is found by adding the first and the last numbers, multiplying this sum by the number of numbers, and then taking half of that product.

Algebraic Novelty 14

Problem: Simple factoring can sometimes also be challenging. Consider the following, where one is asked to factor the following into the product of two binomials with integral coefficients:

$$x^3 + x^2 + x + 1.$$

Solution: We begin by factoring the first two terms as follows: $x^3 + x^2 + x + 1 = x^2(x + 1) + x + 1$. When we then factor the $x + 1$ term, we get $x^3 + x^2 + x + 1 = x^2(x + 1) + x + 1 = (x + 1)(x^2 + 1)$. We have then achieved the desired goal.

Algebraic Novelty 15

Problem: For what positive value of x is $x^4 + 6x^3 + 11x^2 + 3x + 31$ the square of an integer?

Solution: To solve this problem much algebraic manipulation will be required. We seek to find y, when $y^2 = x^4 + 6x^3 + 11x^2 + 3x + 31$. First, we shall factor this expression as:

$$y^2 = x^2(x^2 + 3x + 1) + 3x^3 + 10x^2 + 3x + 31$$
$$= y^2 = x^2(x^2 + 3x + 1) + 3x(x^2 + 3x + 1) + x^2 + 31$$
$$= (x^2 + 3x)(x^2 + 3x + 1) + x^2 + 31.$$

If we can get $(x^2 + 3x + 1) = x^2 + 31$, we would then have $= (x^2 + 3x)(x^2 + 3x + 1) + (x^2 + 3x + 1)$, where we could then factor the $(x^2 + 3x + 1)$ to get $(x^2 + 3x + 1)[x^2 + 3x + 1]$, which is a square number. Therefore, we must find the value of x that enables us to achieve this relationship. Recall that this can only be true if $(x^2 + 3x + 1) = x^2 + 31$.

This easily leads to $3x + 1 = 31$, and therefore, $x = 10$, which solves the original problem by making the algebraic expression a square.

Algebraic Novelty 16

Problem: Find the largest positive integer n less than 100 that is not the square of an integer and where $\sqrt{10 + \sqrt{n}}$ can be expressed as $\sqrt{x} + \sqrt{y}$, when x and y are unequal positive integers.

Solution: We are given that $\sqrt{x} + \sqrt{y} = \sqrt{10 + \sqrt{n}}$. When we square both sides, we get $x + y + 2\sqrt{xy} = 10 + \sqrt{n}$, so that $x + y = 10$, and $2\sqrt{xy} = \sqrt{n}$. Once again squaring both sides, we get $4xy = n$, or $xy = \frac{n}{4}$. We need to maximize the product of the two numbers x and y they need to be as close as possible, but since they cannot be equal, we must use $x = 4$ and $y = 6$ (or the reverse). Therefore, since $(4)(6) = \frac{n}{4}$, we have $n = 96$.

Algebraic Novelty 17

Problem: Find the value of a_4, if $a_n = a_1 a_{n-1} + a_2 a_{n-2} + a3a_{n-3} + \cdots + a_{n-1}a_1$, and $a_n = 0$, when $n \leq 0$, and if $a_1 = 2$.

Solution: Seeking a pattern guides us to approach this problem sequentially. We get

$$a_2 = a_1 a_1 = 2 \cdot 2 = 4$$

$$a_3 = a_1 a_2 + a_2 a_1 = 2 \cdot 4 + 4 \cdot 2 = 8 + 8 = 16$$

$$a_4 = a_1 a_3 + a_2 a_2 + a_3 a_1 = 2 \cdot 16 + 4 \cdot 4 + 16 \cdot 2 = 32 + 16 + 32 = 80.$$

Algebraic Novelty 18

Problem: Solve the following equation for the real value of x:
$\sqrt{\frac{1}{x} + 1} + \frac{1}{x} + 1 = 6.$

Solution: We begin by letting $y = \sqrt{\frac{1}{x}+1}$. We can then rewrite the original equation as follows: $y + y^2 = 6$, or $y^2 + y - 6 = 0 = (y + 3)(y - 2)$. When $y = 2$, then $2 = \sqrt{\frac{1}{x}+1}$, so that $4 = \frac{1}{x}+1$, and $x = \frac{1}{3}$. The other value (when $y = -3$) for $x = -\frac{1}{4}$ will result in an imaginary result, which the original question did not accept.

Algebraic Novelty 19

Problem: The arithmetic mean of two positive numbers $\left(\frac{a+b}{2}\right)$ exceeds the geometric mean of these two numbers $(\sqrt{ab})$ by 50. What is the difference between the square roots of these two numbers?

Solution: We will let the two numbers be represented by a and b, so that we then have $\frac{a+b}{2} - \sqrt{ab} = 50$, or $a + b - 2\sqrt{ab} = 100$. Using some algebraic insight, we then factor the left side of the equation to get $(\sqrt{a} - \sqrt{b})^2 = 100$, then $\sqrt{a} - \sqrt{b} = 10$. This answers the question.

Algebraic Novelty 20

Problem: Find the values of x, y, and z that satisfy the following equations:

$$x^2 y^2 z = 225$$

$$xy^2 z^2 = 75$$

$$x^2 yz^2 = 45.$$

Solution: Curiously, when we multiply these three equations together, we get the following product: $x^5 y^5 z^5 = (5^2 \cdot 3^2)(5^2 \cdot 3)(3^2 \cdot 5) = 3^5 \cdot 5^5$. Therefore, $xyz = 3 \cdot 5 = 15$. By squaring both sides of the equation, we have $x^2 y^2 z^2 = 225$. We now divide each of the original three equations by $x^2 y^2 z^2 = 225$.

$$\frac{x^2 y^2 z^2}{x^2 y^2 z} = \frac{225}{225}, \text{ or } z = 1$$

$$\frac{x^2 y^2 z^2}{x y^2 z^2} = \frac{225}{75} = \frac{3}{1}, \text{ or } x = 3$$

$$\frac{x^2 y^2 z^2}{x^2 y z^2} = \frac{225}{45} = \frac{5}{1}, \text{ or } y = 5.$$

Thus, using some arithmetic acrobatics, we have found the three values of x, y, and z.

Algebraic Novelty 21

Problem: We look to find a positive integer that when its fourth power is subtracted from the square of the sum of the integer and the integer's square, is the same result as one gets by multiplying by 12 the result of taking 3 less than the sum of the positive integer and its square.

Solution: When we translate this into symbols by representing the sought after number as n, we get the following equation: $(n + n^2)^2 - n^4 = 12(n + n^2 - 3)$. This can be then rewritten as $(n + n^2)^2 - 12(n + n^2) + 36 = (n^2 + n - 6)^2 = n^4$. By taking the square root of both sides, we get $n^2 + n - 6 = n^2$, where we see immediately that $n = 6$. Note: When taking the square root of both sides, the $-n^2$ leads to a fraction and a negative number, which does not fit the original question.

Algebraic Novelty 22

Problem: Find all positive integers n for which $n^5 - 1$ is a prime number.

Solution: The term $n^5 - 1$ is factorable so that $n^5 - 1 = (n - 1)(n^4 + n^3 + n^2 + n + 1)$. Therefore, $n^5 - 1$ cannot be a prime number unless $(n - 1)$ becomes 1 so that the factor disappears. This happens only when $n = 2$. We then have $n^5 - 1 = 31$, which is a prime number.

Algebraic Novelty 23

Problem: The challenge here is to find the values of x and y that satisfy these two equations:

$$\frac{x}{x^2 + y^2} = 1 \text{ and } \frac{y}{x^2 + y^2} = 2.$$

Solution: Unexpectedly, when we add these two equations we get the following:

$$\frac{x}{x^2 + y^2} + \frac{y}{x^2 + y^2} = \frac{x + y}{x^2 + y^2} = 3.$$

Then when we divide these two equations, we get the following:

$$\frac{\dfrac{x}{x^2 + y^2}}{\dfrac{y}{x^2 + y^2}} = \frac{x}{y} = \frac{1}{2}.$$

This leads to $y = 2x$. Now substituting for y in the previous sum, we get

$$\frac{x + y}{x^2 + y^2} = \frac{3x}{5x^2} = 3,$$

which leads to the quadratic equation $15x^2 - 3x = 0$ with roots $x = \frac{1}{5}$, and $x = 0$, the latter of which is to be ignored. Therefore, $x = \frac{1}{5}$, and $y = \frac{2}{5}$.

Algebraic Novelty 24

Problem: Consider a polynomial of the form $ax^4 + bx^3 + cx^2 + dx + e$, where the polynomial has integral coefficients whose greatest common divisor is 1, and where $a > 0$. Find the polynomial that would have a zero value when $x = \sqrt{-7}$, and when $x = \sqrt{3}$.

Solution: When $x = \sqrt{-7}$, and $x = \sqrt{3}$, the polynomial must be divisible by $x^2 + 7$ and $x^2 - 3$, or by $(x^2 + 7) \cdot (x^2 - 3) = x^4 + 4x^2 - 21$. Since the coefficients have no common factor greater than 1, and we know that the polynomial is of degree 4 (since a must be positive), the leaves us with the accepted result that $x^4 + 4x^2 - 21$.

Algebraic Novelty 25

Problem: In a school class each student was photographed, gave one of their photographs to every other student, and received one in return from each student. If there were 870 photographs exchanged, how many students were in the class?

Solution: If there were x students in the class, then there were $x(x-1)$ photographs exchanged. Therefore, $x(x-1) = 870$, or $x^2 - x - 870 = 0$, so that $x = 30$ and $x = -29$. We then use the positive answer, which indicates that there were 30 students in the class.

Algebraic Novelty 26

Problem: Find the sum of the finite sequence:

$$\frac{1}{\sqrt{2}+\sqrt{1}} + \frac{1}{\sqrt{3}+\sqrt{3}} + \frac{1}{\sqrt{4}+\sqrt{3}} + \cdots + \frac{1}{\sqrt{25}+\sqrt{24}}.$$

Solution: In order to find the sum, we need multiply each fraction by the unit fraction formed by the denominator's conjugate in both numerator and denominator. That enables us to rationalize the denominators as follows:

$$\frac{1}{\sqrt{2}+\sqrt{1}} \cdot \frac{\sqrt{2}-\sqrt{1}}{\sqrt{2}-\sqrt{1}} + \frac{1}{\sqrt{3}+\sqrt{2}} \cdot \frac{\sqrt{3}-\sqrt{2}}{\sqrt{3}-\sqrt{2}} + \frac{1}{\sqrt{4}+\sqrt{3}} \cdot \frac{\sqrt{4}-\sqrt{3}}{\sqrt{4}-\sqrt{3}} + \cdots + \frac{1}{\sqrt{25}+\sqrt{24}} \cdot \frac{\sqrt{25}-\sqrt{24}}{\sqrt{25}-\sqrt{24}}$$

$$= \frac{\sqrt{2}-\sqrt{1}}{2-1} + \frac{\sqrt{3}-\sqrt{2}}{3-2} + \frac{\sqrt{4}-\sqrt{3}}{4-3} + \cdots + \frac{\sqrt{25}-\sqrt{24}}{25-24} = \sqrt{25} - \sqrt{1} = 5 - 1 = 4.$$

Algebraic Novelty 27

Problem: Given that $(x^2 - x - 1)$ and $(ax - 1)$ are factors of $ax^3 + bx^2 + 1$, and that a and b are integers, what is the value of b?

Solution: We can begin by multiplying the factors:

$$ax^3 + bx^2 + 1 = (ax - 1)(x^2 - x - 1) = ax^3 + (-a - 1)x^2 + (1 - a)x + 1.$$

By equating coefficients of x^2 and x we find that $b = -a - 1$, and $1 - a = 0$. Therefore, $a = 1$, and $b = -2$.

Algebraic Novelty 28

Problem: Given that $\sin x = 3 \cos x$, what is $\sin x \cos x$ equal to?

Solution: Multiply the given equation first by $\sin x$ to get $\sin^2 x = 3 \sin x \cos x$, and then by $\cos x$ to get $\cos^2 x = (\frac{1}{3})\sin x \cos x$. We then add these two new equations so that $\sin^2 x + \cos^2 x = (3 + \frac{1}{3})\sin x \cos x = \frac{10}{3}\sin x \cos x$. However, by the Pythagorean theorem we know that $\sin^2 x + \cos^2 x = 1$; therefore, $\frac{10}{3}\sin x \cos x = 1$, and $\sin x \cos x = \frac{3}{10}$.

Algebraic Novelty 29

Problem: Given that $\frac{b}{a} = 2$, and $\frac{c}{b} = 3$, what is the value of $\frac{a+b}{b+c}$?

Solution: Since we know that $\frac{c}{b} = 3$, we can get $c = 3b$; also, since $\frac{b}{a} = 2$, we have $b = 2a$. By substitution we get $c = 3(2a) = 6a$. We are now ready to set up the fraction by making the appropriate replacements shown here:

$$\frac{a+b}{b+c} = \frac{a+2a}{2a+6a} = \frac{3}{8}.$$

Algebraic Novelty 30

Problem: In the equation that follows, the values of a and b are constants. The challenge is to find the actual values of a and b so that the following is true: $(x + 2)(x + a) = x^2 + bx + 6$.

Solution: Multiplying the two binomials on the left side of the equation and setting it equal to the right side of the equation yields $x^2 + (a + 2)x + 2a = x^2 + bx + 6$. We can now equate the coefficients of x and the constants. So that $a + 2 = b$, and $2a = 6$. Therefore, $a = 3$, and $b = 5$.

Algebraic Novelty 31

Problem: We seek the largest positive integer n for which $n^3 + 100$ is divisible by $n + 100$.

Solution: We begin by noting that

$$n^3 + 100 = (n + 100)(n^2 - 10n + 100) - 900.$$

We also know that if $n + 10$ divides $n^3 + 100$, then it must have 900 as a factor. Since n is at its largest whenever $n + 1$ is maximized, and since the largest divisor of 900 is actually 900, we find that $n + 1 = 900$, and therefore, $n = 890$.

Algebraic Novelty 32

Problem: Solve the following equation for x:

$$\sqrt{\frac{3x+2}{6x-4}} + \sqrt{\frac{6x-4}{3x+2}} = \frac{5}{2}.$$

Solution: To simplify matters a bit, we begin by letting

$$\sqrt{a} + \sqrt{\frac{1}{a}} = \frac{5}{2}, \text{ or } \sqrt{a} + \frac{1}{\sqrt{a}} = \frac{5}{2}, \text{ so that } \frac{a+1}{\sqrt{a}} = \frac{5}{2}.$$

Then by cross multiplying we get $5\sqrt{a} = 2(a+1)$, and then squaring both sides of this last equation we have $25a = 4(a^2 + 2a + 1)$. Simplifying this we get $4a^2 - 17a + 4 = 0$, or $(4a - 1)(a - 4) = 0$, so that $a = \frac{1}{4}$ and $a = 4$. When $a = \frac{1}{4}$ we have

$$\frac{3x+2}{6x-4} = \frac{1}{4}, \text{ or } 12x + 8 = 6x - 4, \text{ and } x = -2.$$

Similarly, when $a = 4$,

$$\frac{3x+2}{6x-4} = 4, \text{ or } 3x+2 = 24x-16, \text{ and } x = \frac{6}{7}.$$

Algebraic Novelty 33

Problem: Find the product for the following four terms:

$$\left(\sqrt{5}+\sqrt{6}+\sqrt{7}\right)\left(\sqrt{5}+\sqrt{6}-\sqrt{7}\right)\left(\sqrt{5}-\sqrt{6}+\sqrt{7}\right)\left(-\sqrt{5}+\sqrt{6}+\sqrt{7}\right).$$

Solution: The solution to this challenge is to recognize the popular algebraic identity $(x+y)(x-y) = x^2 - y^2$. We then apply it to the first two parenthetical terms and then to the second two terms to get the following:

$$\left(\sqrt{5}+\sqrt{6}+\sqrt{7}\right)\left(\sqrt{5}+\sqrt{6}-\sqrt{7}\right) = \left(\sqrt{5}+\sqrt{6}\right)^2 - \left(\sqrt{7}\right)^2 = \left(11+2\sqrt{30}\right)-7 = 4+2\sqrt{30}.$$

$$\left(\sqrt{5}-\sqrt{6}+\sqrt{7}\right)\left(-\sqrt{5}+\sqrt{6}+\sqrt{7}\right) = \left(\sqrt{7}\right)^2 - \left(\sqrt{6}+\sqrt{7}\right)^2 = 7 - \left(11-2\sqrt{30}\right) = -4+2\sqrt{30}.$$

We now simply have to multiply the last two expressions to get $(4+2\sqrt{30})(-4+2\sqrt{30}) = (2\sqrt{30})-4^2 = 120-16 = 104$, which is the product of the four given terms.

Algebraic Novelty 34

Problem: What is the sum of the solutions for the equation

$$\sqrt[4]{x} = \frac{12}{7-\sqrt[4]{x}}?$$

Solution: To simplify the situation, begin by letting $y = \sqrt[4]{x}$ so that the original equation can be written as $y = \frac{12}{7-y}$ or $y^2 - 7y + 12 = 0$, whose roots are $y = 3$, and $y = 4$. However, we seek the values for x. Since $\sqrt[4]{x} = 3$, then $x = 3^4$, then $x = 3^4 = 81$, and similarly, $x = 4^4 = 256$, the sum of the solutions is $81 + 256 = 337$.

Algebraic Novelty 35

Problem: The results of a mathematics competition showed that the sum of the scores of Barbara and David equal the sum of the scores of Alfred and Charlie. If the scores of Barbara and Charlie had been interchanged, then the sum of the scores of Alfred and Charlie would have exceeded the sum of the scores of Barbara and David. Furthermore, David's score exceeded the sum of the scores of Barbara and Charlie. The challenge is to determine the order in which the four contestants finished from highest to lowest scores. (All scores are non-negative.)

Solution: To simplify the problem solution we let A, B, C, and D represent the scores of Alfred, Barbara, Charlie, and David, respectively. Therefore, we get:

$$A + C = B + D \tag{1}$$

$$A + B > C + D \tag{2}$$

$$D > B + C \tag{3}$$

By adding (1) and (2), we get $2A > 2D$ or $A > D$. By subtracting (2) − (1), we get $B - C > C - B$, or $B > C$. Then from (3) we have $D > B$ (since C is not negative).

This results in the following: $A > D > B > C$.

Algebraic Novelty 36

Problem: Consider ordered triples of the form (x, y, z) consisting of nonzero real numbers that have the property that each number is the product of the other two numbers. What are these triples?

Solution: The property described above is such that $x = yz$, $y = xz$, and $z = xy$. This could give us the following: $xyz = (yz)(xz)(xy) = (xyz)^2$. We cannot have $xyz = 0$, since we ruled that out at the start. Therefore, the only other way that the above relationship can hold is if $xyz = 1$.

We can also look at this as follows: since $x = yz$, we have $x^2 = xyz$. Similarly, $xyz = x^2 = y^2 = z^2$, so that the actual values $x = y = z = 1$. It is not possible for all three values x, y, and z to be -1 since then it would not be possible to have $x = yz$. This leaves the only possibilities for (x, y, z) to be $(1,1,1)$, $(-1,-1,1)$, $(1,-1,-1)$, $(-1,1,-1)$.

Algebraic Novelty 37

Problem: Given that $xy = 6$ and $x^2y + xy^2 + x + y = 63$, find the value of $x^2 + y^2$.

Solution: This can be most easily solved by factoring the given expression as follows:

$$63 = x^2y + xy^2 + x + y = xy(x + y) + (x + y) = (xy + 1)(x + y).$$

Since $xy = 6$, we then get $(6 + 1)(x + y) = 63$, which then becomes $x + y = 9$. Since we seek $x^2 + y^2$, we know that $(x + y)^2 = x^2 + y^2 + 2xy$, so then $x^2 + y^2 = (x + y)^2 - 2xy = 81 - 12 = 69$.

Algebraic Novelty 38

Problem: Given that a and b are each positive integers greater than 1, which of the following has the greatest value?

$$a + b$$
$$a - b$$
$$\sqrt{2ab}$$
$$\frac{a^2 + b^2}{a + b}$$
$$\frac{a^4 + b^4}{a^3 + b^3}.$$

Solution: At first glance, this appears to be an overwhelming algebraic challenge. However, we might want to see if there is a pattern or some other hidden aspect to be found. One way to approach this is to replace the letters with numbers. For example, let $a = 4$ and $b = 2$. We then get the following results:

$$a + b = 4 + 2 = 6$$
$$a - b = 4 - 2 = 2$$
$$\sqrt{2ab} = \sqrt{16} = 4$$
$$\frac{a^2 + b^2}{a + b} = \frac{16 + 4}{4 + 2} = 3.333\overline{3}$$
$$\frac{a^4 + b^4}{a^3 + b^3} = \frac{256 + 16}{64 + 8} = 3.777\overline{7}.$$

Solving the simplified version enabled us to readily solve the original problem, where $a + b$ has the greatest value.

Algebraic Novelty 39

Problem: What is the units digit of 7^{23}?

Solution: This problem appears, on the surface, to be rather complex since the number is quite large. It is even too large to fit onto a calculator display. With a number that large, there may be a pattern which can help us arrive at an answer. Suppose we take increasingly larger powers of 7, starting with 7^1, to see if a pattern emerges, which would then enable us to apply it to our original problem.

$$7^1 = \underline{7} \qquad 7^5 = 1690\underline{7}$$
$$7^2 = 4\underline{9} \qquad 7^6 = 11764\underline{9}$$
$$7^3 = 34\underline{3} \qquad 7^7 = 82354\underline{3}$$
$$7^4 = 240\underline{1} \qquad 7^8 = 576480\underline{1}$$

We notice that the units digits repeat in the cycles of 4. Thus, we can now return to our original problem. Since the exponent 23 (of our original problem 7^{23}) has a remainder of 3 when divided by 4, we can assume that its terminal digit is the same as that of 7^{19}, 7^{15}, 7^{11}, and 7^7, which we now know is 3. Once again, turning to a simpler version of the problem enables us to solve the original problem in a relatively simple manner.

Algebraic Novelty 40

Problem: Show that the equation $x^4 - 5x^3 - 4x^2 - 7x + 4 = 0$ has no negative roots.

Solution: This may appear to be somewhat complicated, but with a bit of clever manipulation, we can determine that it has no negative roots without even solving the equation. To do this, we rewrite this equation in the following form: $5x^3 + 7x = x^4 - 4x^2 + 4$, which is then $5x^3 + 7x = (x^2 - 2)^2$. Suppose we consider a negative value for x, then the left side of the equation will be negative, but the right side will always be positive. Therefore, no negative value can satisfy this equation.

Algebraic Novelty 41

Problem: Find the value of $x + y$, if $x^2 + y^2 = 16$ and $xy = -10$.

Solution: Most people's initial reaction to this problem is to attempt to solve the two given equations simultaneously. After all, we are given two equations and two variables; the simultaneous solution should lead to values for x and y. However, this proves to be a rather tedious process. Furthermore, we notice that neither equation is linear: one represents a graph of a circle, while the other yields a rectangular hyperbola. There might be as many as four sets of answers. Let's consider a different point of view. One might ask why the problem is calling for "the value of $x + y$", and not the individual values of x and y. How are x^2 and y^2 and xy related? Where might one have seen these two expressions together? Aha! Both appear in the expansion of $(x + y)^2$.

Consider $(x + y)^2 = x^2 + y^2 + 2xy$. We are given that $x^2 + y^2 = 16$ and $xy = -10$. Then making proper replacements, we get $(x + y)^2 = x^2 + y^2 + 2xy = 36 + (-20) = 16$.

So, if $(x + y)^2 = 16$, then it follows that $x + y = \pm 4$, which we arrived at by looking at the problem from a different point of view.

Algebraic Novelty 42

Problem: The sum of two numbers is 2. The product of the same two numbers is 5. Find the sum of the reciprocals of these two numbers.

Solution: Most people upon seeing the simple question will immediately generate the equations $x + y = 2$, and $xy = 5$. When facing two equations with two unknowns, most people recall that they were taught to solve these equations simultaneously by substitution.

These two equations can be solved simultaneously using the quadratic formula, which is

$$x = \frac{-b \pm \sqrt{b^2 - 4ac}}{2a}, \text{ for } ax^2 + bx + c = 0.$$

However, the method yields complex values for both x and y, namely, $1 + 2i$, and $1 - 2i$. Following the requirements of the original problem, we now need to take the sum of the reciprocals of these two roots:

$$\frac{1}{1+2i} + \frac{1}{1+2i} = \frac{(1-2i)+(1+2i)}{(1+2i)(1-2i)} = \frac{2}{5}.$$

We should emphasize here that there is nothing wrong with this method, it is just not the most elegant way to solve this problem. A more efficient and elegant solution would be to step back from the problem and see what the question requires. Curiously, this problem is not asking for the values of x and y, but rather the sum of the reciprocals of these two numbers. That is, we seek to find $\frac{1}{x} + \frac{1}{y}$. Using a strategy of working backwards, we could ask ourselves from what might this have come. Adding these two fractions could give us his answer. Therefore, $\frac{1}{x} + \frac{1}{y} = \frac{x+y}{xy}$. At this point the required answer is immediately available to us since we know the sum of the numbers is 2, and the product of the numbers is 5. We merely substitute these values in the last fraction to get $\frac{1}{x} + \frac{1}{y} = \frac{x+y}{xy} = \frac{2}{5}$, and our problem is solved.

Algebraic Novelty 43

Problem: Find the sum of the first 30 odd numbers.

Solution: A student with experience in elementary algebra will typically employ the formula to find the sum of the arithmetic sequence $S = \frac{n}{2}(f + l)$, where n represents the number of items, f is the first number, and l is the last number. However, in order to use the formula, one needs to know what the last number would be. With a little cleverness, we can arrive at the fact that 59 is the last number of this arithmetic sequence. Therefore, applying the formula we get: $S = \frac{30}{2}(1 + 59) = 900$.

Another clever way to solve this problem is to inspect the pattern that evolves from the sum of consecutive odd numbers. (See Figure 3-1.)

Consecutive odd numbers	Number of odd numbers	Sums of consecutive odd numbers
1	1	1
$1 + 3$	2	4
$1 + 3 + 5$	3	9
$1 + 3 + 5 + 7$	4	16
$1 + 3 + 5 + 7 + 9$	5	25
. . .		
$1 + 3 + 5 + 7 + 9 + \cdots + n$	n	n^2

Figure 3-1

The pattern that evolves is that the sum of consecutive odd numbers, beginning with 1, is the square of the number of odd numbers in the sequence, which in our original problem was $30^2 = 900$.

Algebraic Novelty 44

Problem: Find the sum of this series:

$$\frac{1}{1 \cdot 2} + \frac{1}{2 \cdot 3} + \frac{1}{3 \cdot 4} + \cdots + \frac{1}{99 \cdot 100}.$$

Solution: One can compute the individual values for each of the fractions and then add them. That is, $\frac{1}{1 \cdot 2} = \frac{1}{2}, \frac{1}{2 \cdot 3} = \frac{1}{6}, \frac{1}{3 \cdot 4} = \frac{1}{12}, \frac{1}{4 \cdot 5} = \frac{1}{20}$, etc. Thus, the final answer would be the sum of $\frac{1}{2} + \frac{1}{6} + \frac{1}{12} + \frac{1}{20} + \cdots + \frac{1}{9,900}$.

Obviously, this would be a rather laborious task, even using a calculator. Let's see if there is a pattern we can use.

$$\frac{1}{2}$$

$$\frac{1}{2}+\frac{1}{6}=\frac{2}{3}$$

$$\frac{1}{2}+\frac{1}{6}+\frac{1}{12}=\frac{3}{4}$$

$$\frac{1}{2}+\frac{1}{6}+\frac{1}{12}+\frac{1}{20}=\frac{4}{5}$$

$$\vdots$$

$$\frac{1}{2}+\frac{1}{6}+\frac{1}{12}+\frac{1}{20}+\cdots+\frac{1}{9,900}=\frac{99}{100}$$

Notice that the last fraction in the sum is the product of two consecutive numbers. This allows us to reach the sum as shown above.

Algebraic Novelty 45

Problem: An airplane flew from A to B, a distance of 450 miles, at an average speed of 180 miles per hour. The return journey was made at an average speed of 90 miles per hour. What was the average speed of the whole journey in miles per hour?

Solution: We must first understand what is meant by "average speed." In problems involving travel at different speeds, it means the single speed which, if maintained throughout the journey, would accomplish the trip in the same total time. For example, if a trip of 120 miles requires 3 hours, the average speed is 40 miles per hour, regardless of how the speed varies during the trip. A little experimentation will show that you can average different speeds if they are maintained for equal periods of time, but *not* if they are maintained for equal distances. There are various ways of calculating the average speed in problems of the kind presented here.

Method 1: Let us use the relationship: Rate × Time = Distance, or Time = $\frac{\text{Distance}}{\text{Rate}}$ to calculate the time for each part of the trip; once we find the total time, we can calculate the average rate over the total distance:

$$T_1 = \frac{D_1}{R_1} = \frac{450 \text{ miles}}{180 \, \frac{\text{miles}}{\text{hour}}} = 2\frac{1}{2} \text{ hours} \qquad T_2 = \frac{D_2}{R_2} = \frac{450 \text{ miles}}{90 \, \frac{\text{miles}}{\text{hour}}} = 5 \text{ hours.}$$

The total time T for the total distance D of 900 miles is $7\frac{1}{2}$ hours. The average rate R is:

$$R = \frac{D}{T} = \frac{900 \text{ miles}}{7\frac{1}{2} \text{ hours}} = 120 \text{ miles per hour.}$$

Method 2: If you like to work with general formulas, instead of working out the arithmetic for each problem, the following approach will appeal to you. Let us find the average rate R for any round trip that is made one way at rate R_1, and the other way at rate R_2. As in Method 1:

$$T_1 = \frac{d}{R_1} \qquad T_2 = \frac{d}{R_2},$$

where d represents the distance going one way.

$$T = T_1 + T_2 = \frac{d}{R_1} + \frac{d}{R_2} = d\left(\frac{1}{R_1} + \frac{1}{R_2}\right).$$

Since the total distance is $2d$:

$$R = \frac{2d}{T} = \frac{2d}{d\left(\frac{1}{R_1} + \frac{1}{R_2}\right)} = \frac{2}{\frac{1}{R_1} + \frac{1}{R_2}} = \frac{2R_1 R_2}{R_1 + R_2}$$

Notice that this expression is the reciprocal of the average of the reciprocals of the two rates. That is, if we take the reciprocals of R_1 and R_2, then find the average of these reciprocals, and then take the reciprocal of this average, we will have the average rate. This quantity is called the *harmonic mean* of two quantities; the harmonic mean for a and b is

$$\frac{1}{\frac{\frac{1}{a}+\frac{1}{b}}{2}}=\frac{2ab}{a+b}.$$

The average rate when half a trip is made at one speed and the other half at another speed is the harmonic mean of the two speeds. In this problem:

$$R=\frac{2\cdot90\cdot180}{90+180}=120.$$

Method 3: The idea behind the harmonic mean method is that the time for a given trip is inversely proportional to the rate. Therefore, in finding the average rate, the slower speed must be given proportionately more "weight." In this problem, we can see that for each hour spent traveling at 180 miles per hour, 2 hours must be spent at 90 miles per hour to cover the same distance. Therefore, out of every 3 hours, there are 2 hours spent at 90 miles per hour and 1 hour spent at 180 miles per hour. We can find the average by adding $90 + 90 + 180$ and dividing by 3: $\frac{90+90+180}{3}=\frac{360}{3}=120$. This is somewhat of an intuitive method and should not be risked unless you fully understand the principles behind it. But if you can use it, it could be the fastest of the three approaches.

Algebraic Novelty 46

Problem: Since Henry received a speeding ticket on his way to work one day, he decided to cut his speed by $16\frac{2}{3}\%$. How much more time will Henry require to get to work with his revised speed?

Solution: Once again, we will use convenient variables as follows: Let r = Henry's original rate, let t = Henry's original travel time, and let d = the travel distance.

It is well known that the product of rate in time equals the distance, we have $r \cdot t = d$. Since Henry's rate is being decreased by $\frac{1}{6}$ (since $16\frac{2}{3}\% - \frac{1}{6}$) his new rate is $\frac{5}{6}$ of his original rate. Therefore, his new time must be $\frac{6}{5}$ his original time, since $\left(\frac{5}{6}r\right)\left(\frac{6}{5}t\right) = d$. However, $\frac{6}{5}t$ represents an increase of t by $\frac{1}{5}$, or a 20% increase in time.

Algebraic Novelty 47

Problem: When checking the cash register, a clerk counts Q quarters, D dimes, N nickels, and P pennies. Later, he discovers that x of the nickels were counted as quarters and x of the dimes were counted as pennies. How can the clerk correct the total obtained?

Solution: The amount of cents first counted by the clerk was:

$$25Q + 10D + 5N + P. \tag{1}$$

Later, he found that x nickels were counted as quarters. Therefore, his count then is $(Q - x)$ quarters and $(N + x)$ nickels. He also found that x of the dimes were counted as pennies, making his count $(P - x)$ pennies, and $(D + x)$ dimes.

Therefore, amount of cents the clerk *actually* has is:

$$25(Q - x) + 5(N + x) + 10(D + x) + (P - x). \tag{2}$$

This can be expanded to:

$$25Q - 25x + 5N + 5x + 10D + 10x + P - x$$
$$= 25Q - 11x + 5N + 10D + P. \tag{3}$$

In order to determine the difference between his original count and his corrected count, we subtract line (1) from line (3) to obtain $-11x$. This means that to obtain the true count from the original erroneous count, we must add $-11x¢$ to the original count. That is, we must subtract $11x¢$.

Algebraic Novelty 48

Problem: Seven pounds of peaches cost as much as 1 pound of apples and 2 pounds of grapes. Seven pounds of apples cost as much as 10 pounds of grapes and 1 pound of peaches. How many pounds of grapes may be purchased for the amount of money required to purchase 12 pounds of apples?

Solution: A logical beginning would be to introduce the following variables:

- Let p = price of one pound of peaches
- Let g = price of one pound of grapes
- Let a = price of one pound of apples

From the first sentence we obtain the relationship:

$$7p = a + 2g. \tag{1}$$

From the second sentence we obtain the relationship:

$$7a = 10g + p. \tag{2}$$

Since there are three variables and only two equations, we know we cannot simply solve for any of the variables. But we can eliminate one variable and obtain a ratio between the other two. In fact, the problem requires us only to find the ratio between g and a; so, our objective is to eliminate p from equations (1) and (2). There are various algebraic techniques for doing this. We can solve equation (1) for p as follows: $p = \frac{a+2g}{7}$. Then by substituting this value into equation (2), we get $7a = 10g + \frac{a+2g}{7}$, which is $48a = 72g$.

Alternatively, we could multiply equation (2) by 7 (to give us $7p$ in both equations) and then add the equations: $(7p = a + 2g) + (49a = 70g + 7p) = (7p + 49a = a + 2g + 70g + 7p)$.

By either method we obtain: $48a = 72g$. This result tells us that 48 pounds of apples cost as much as 72 pounds of grapes. The problem asks for the equivalent of 12 pounds of apples. This equation is equivalent to $12a = 18g$. Therefore, the answer is 18 pounds of grapes.

Algebraic Novelty 49

Problem: Simplify each of the following without a calculator:

a) $\dfrac{729^{35}-81^{52}}{27^{69}}$,

b) $\dfrac{6\cdot27^{12}+2\cdot81^{9}}{8000000^{2}}\cdot\dfrac{80\cdot32^{3}\cdot125^{4}}{9^{19}-729^{6}}$.

Solution: Although one may be tempted to use a calculator to evaluate his expression, all too often we overestimate a calculator's abilities only for it to come back with an "error" message. With knowledge of the powers of 3, the first expression can be rather cleverly evaluated as follows:

a) $\dfrac{729^{35}-81^{52}}{27^{69}}=\dfrac{(3^{6})^{35}-(3^{4})^{52}}{(3^{3})^{69}}=\dfrac{3^{210}-3^{208}}{3^{207}}=\dfrac{3^{208}\cdot(3^{2}-1)}{3^{207}}=3\cdot8=24.$

The second expression can be simplified by breaking the numbers up into prime factors as follows:

b) $\dfrac{6\cdot27^{12}+2\cdot81^{9}}{8000000^{2}}\cdot\dfrac{80\cdot32^{3}\cdot125^{4}}{9^{19}-729^{6}}=\dfrac{2\cdot3\cdot(3^{3})^{12}+2\cdot(3^{4})^{9}}{(3^{2})^{19}-(3^{6})^{6}}\cdot\dfrac{2^{4}\cdot5\cdot(2^{5})^{3}\cdot(5^{3})^{4}}{(2^{3}\cdot2^{6}\cdot5^{6})^{2}}$

$=\dfrac{2\cdot3^{37}+2\cdot3^{36}}{3^{38}-3^{36}}\cdot\dfrac{2^{19}\cdot5^{13}}{2^{18}\cdot5^{12}}=\dfrac{2\cdot3^{36}(3+1)}{3^{36}(3^{2}-1)}\cdot2\cdot5=\dfrac{2(3+1)}{3^{2}-1}\cdot2\cdot5=10.$

Algebraic Novelty 50

Problem: Evelyn, Henry, and Al are playing a game. Evelyn loses Round 1 and gives Henry and Al as much money as they each then have. In Round 2, Henry loses and gives Evelyn and Al as much money as they each then have. Al loses in Round 3 and gives Evelyn and Henry as much money as they each have. They decide to quit at this point and discover that they each have $24. How much money did each person start with?

Solution: Begin this problem by setting up a system of three equations with three variables, as we show in Figure 3-2, and simulating the process as it is described above.

Round	Evelyn	Henry	Al
Start	x	y	z
2	$x - y - z$	$2y$	$2z$
3	$2 - 2y - 2x$	$3y - x - z$	$4z$
4	$4x - 4y - z$	$6y - 2x - 2z$	$7z - x - y$

Figure 3-2

The table leads us to this system of equations, where each person had 24 dollars:

$$4x - 4y - 4z = 24,$$

$$-2x + 6y - 2z = 24,$$

$$-x - y - 7z = 24.$$

Solving the system leads to $x = 39$ $y = 21$ and $z = 12$. Thus, Evelyn began with $39, Henry began with $21, and Al began with $12.

The problem stated the situation at the end of the story ("They each have $24") and asked for the starting situation ("How much money did they each start with?"). This is almost a sure sign that a working backwards strategy could be employed. Let's see how this makes our work easier. In Figure 3-3, we begin at the end with each having $24.

Round	Evelyn	Henry	Al
End of Round 3	24	24	24
End of Round 2	12	12	48
End of Round 1	6	42	24
Start	39	21	12

Figure 3-3

Evelyn started with $39, Henry with $21, and Al with $12—the same answers we arrived at by solving the problem algebraically.

Algebraic Novelty 51

Problem: Nancy breeds New Zealand rabbits as a hobby. During April, the number of rabbits increased by 10%. In May, 10 new rabbits were born, and at the end of May, Nancy sold one third of her flock. During June, 20 new rabbits were born, and at the end of June, Nancy sold one half of her total flock. So far in July, 5 rabbits have been born, and Nancy now has 55 rabbits. How many rabbits did Nancy start with on April 1?

Solution: We can begin with an algebraic approach. Let x represent the number of rabbits Nancy started with on April 1. Then,

$$x + \frac{x}{10} = \frac{11x}{10} = \text{no. of rabbits at the end of April}$$

$$\left(\frac{2}{3}\right)\left(\frac{11x}{10} + 10\right) = \frac{22x}{30} + \frac{20}{3} = \text{no. of rabbits at the end of May}^2$$

$$\left(\frac{1}{2}\right)\left(\frac{22x}{30} + \frac{20}{3} + 20\right) = \text{no. of rabbits at the end of June}$$

$$\left(\frac{1}{2}\right)\left(\frac{22x}{30} + \frac{20}{3} + \frac{20}{1}\right) + 5 = 55 = \text{Nancy's total rabbits in July}$$

$$\left(\frac{1}{2}\right)\left(\frac{22x}{30} + \frac{20}{3} + 20\right) = 50$$

$$\left(\frac{22x}{30} + \frac{20}{3} + 20\right) = 100$$

$$\frac{22x}{30} + \frac{20}{3} = 80$$

$$22x = 2200$$

$$x = 100.$$

Thus, Nancy started with 100 rabbits.

Notice that the problem tells us how many rabbits Nancy had at the end of the situation, and we are asked to find how many she began with.

Let's apply our *working backwards* strategy. We perform the inverse operations consecutively, working backwards, as shown in Figure 3-4.

$55 - 5 = 50$	Start of July
$50 \cdot 2 = 100$	June: sold $\frac{1}{2}$, therefore, must have been 100
$100 - 20 = 80$	20 were born in June – must have been 80
120	Sold $\frac{1}{3}$, thus, the 80 represents the $\frac{2}{3}$ she had, so we multiply by $\frac{3}{2}$
110	10 were born in May, so we subtract 10
100	To get to the start, we need to find the number, which when increased by 10% is 110; that number is 100

Figure 3-4

Therefore, Nancy started with 100 rabbits.

Algebraic Novelty 52

Problem: Which is larger: $\sqrt[9]{9!}$ or $\sqrt[10]{10!}$?

Solution: Using an algebraic method to prove the general case for this situation is to prove that $\sqrt[n+1]{(n+1)!} > \sqrt[n]{n!}$. Although a bit tedious, we will provide this proof here. We can then better compare it to the problem-solving strategy we shall offer as an alternative.

Since $\sqrt[n+1]{n+1} > \sqrt[n+1]{n}$, $\sqrt[n+1]{n!(n+1)} > \sqrt[n+1]{n!n}$ (multiplying by $\sqrt[n+1]{n!}$); that is, $\sqrt[n+1]{(n+1)!} > \sqrt[n+1]{n!n}$. Since $n^n > n!$ for $n > 1$, we have $n > \sqrt[n]{n!}$. Therefore, $n!n > n!(n!)^{\frac{1}{n}} = (n!)^{\frac{n+1}{n}}$, and $\sqrt[n+1]{n!n} > \sqrt[n]{n!}$. Thus, $\sqrt[n+1]{(n+1)!} > \sqrt[n+1]{n!n} > \sqrt[n]{n!}$.

As an alternative to this rather complicated proof, we can approach this problem by working backwards. That is, we begin with what we want to establish. Let us start by taking both terms to a common power, namely, the 90th power: $(\sqrt[9]{9!})^{90} <? (\sqrt[10]{10!})^{90}$. Then $(9!)^{10} <? (10!)^{9}$, which can be written as $(9!)^{9}(9!) <? (9!)^{9}(10!)^{9}$, or $9! < 10^{9}$. Therefore, $\sqrt[9]{9!} < \sqrt[10]{10!}$.

Algebraic Novelty 53

Problem: Two positive integers differ by 5. If their square roots are added, the sum is also 5. What are the two integers?

Solution: The traditional approach is to set up a system of equations as follows:

Let x = the first integer, and let y = the second integer. Then, $y = x + 5$, and $\sqrt{x} + \sqrt{y} = 5$. By substituting for y, we get $\sqrt{x} + \sqrt{x+5} = 5$. Squaring both sides, $x + x + 5 + 2\sqrt{x(x+5)} = 25$, then simplifying gives us $2\sqrt{x(x+5)} = -2x + 20$. Squaring again, we have $4x^2 + 20x = 4x^2 - 80x + 400$, or $100x = 400$, so that $x = 4$, and $y = 9$. Thus, the two integers are 4 and 9.

Obviously, this procedure requires a knowledge of radicals and a great deal of careful algebraic manipulation. As an alternative, let us make use of some intelligent guessing and testing to solve this problem. Since the sum of the square roots of the two integers is 5, the individual square roots must be 4 and 1, or 3 and 2. Thus the integers must be 16 and 1, or 9 and 4. However, only 9 and 4 have a difference of 5, and must, therefore, be the correct answer.

Algebraic Novelty 54

Problem: Find all the integral values of x which satisfy the equation $(3x+7)^{(x^2-9)} = 1$.

Solution: A linear expression raised to a quadratic exponent appears to require both advanced algebraic facility and a lengthy, complex solution. Instead, let's examine a simpler, analogous version of the problem to find out what is really taking place here. For example, let's look at $a^b = 1$. This problem is a bit less frightening, and somewhat easier to examine and discuss. The expression will have a value of 1 when the base a is 1 since $(1)^b = 1$ for any value of b. Similarly, the expression will also have the value 1 when the exponent is 0, since $(a)^0 = 1$ for any non-zero value of a. We now go back to the original problem and apply what we have found in our examination of the simpler version.

Case 1: Since 1 raised to any power equals 1, set the base equal to 1, so that $3x + 7 = 1$, and $x = -2$.

Case 2: Since any non-zero expression raised to the 0 power equals 1, set the exponent equal to 0, so that $x^2 - 9 = 0$, and $(x - 3)(x + 3) = 0$, so that $x = 3$ and $x = -3$.

Case 3: When -1 is raised to an even power, it also has a value of $+1$. Consider $3x + 7 = -1$, then $x = -\frac{8}{3}$, which is non-integral.

Case 4: When 0 is raised to the 0 power the value is 1. But because $3x + 7 = 0$ gives us $x = -\frac{7}{3}$, and $x^2 - 9 = 0$ yields $x = \pm 3$, this case is impossible here.

Thus, there are three integral values of x for which the equation is correct, namely, $+3$, -3, and -2.

Algebraic Novelty 55

Problem: Find the value of the following expression when $x = 6$:

$$(x - 10)(x - 9)(x - 8) \cdots (x - 3)(x - 2)(x - 1).$$

Solution: One can write out the ten terms in the sequence, substitute 6 for x in each term, and then multiply the resulting 10 numbers. Instead, let's examine this problem from another point of view. Within the ten-term sequence, we find $(x - 6)$ as one of the factors. Since this factor will have the value 0 when we substitute 6 for x, the entire expression will have the value 0. Thus, the problem is easily solved. It sometimes pays to step back from the problem and consider various aspects before plowing into the traditional procedures.

Algebraic Novelty 56

Problem: Find the value of $(x + y)$ if $123x + 321y = 345$, and $321x + 123y = 543$.

Solution: The typical algebraic solution here would probably be far too complicated, so one needs to look for a trick solution. By adding these two equations we get 444x + 444y = 888, and by dividing by 444, gives us $x + y = 2$. Problem solved!

Algebraic Novelty 57

Problem: If $\frac{1}{x+5} = 4$, what is the value of $\frac{1}{x+6}$?

Solution: Traditionally, when faced with a problem of this type, one will seek to find the value of x by solving the original equation. If we solve this equation for x, we obtain $x = -\frac{19}{4}$.

Then, we substitute for x in the fraction expression $\frac{1}{x+6}$ and obtain $\frac{4}{5}$. Of course, this will involve some rather cumbersome algebraic and arithmetic manipulations. However, using some logical reasoning, we can approach the problem differently. If $\frac{1}{x+5} = 4$, then we can take the reciprocals of both sides and obtain $x + 5 = \frac{1}{4}$. We now add 1 to both sides, giving us $x + 6 = \frac{5}{4}$. We again take the reciprocals of both sides to obtain $\frac{1}{x+6} = \frac{4}{5}$. This demonstrates the beauty of carefully inspecting and algebraic challenge before delving into it.

Algebraic Novelty 58

Problem: Given that $\frac{1}{x+y} = \frac{1}{x} + \frac{1}{y}$ we need to find the value of $\left(\frac{x}{y}\right)^3$.

Solution: We begin by simplifying fractions by multiplying both sides of the equation by the common denominator $xy(x + y)$ of the fractions to get:

$$xy = y(x+y) + x(x+y) = x^2 + y^2 + 2xy,$$

so that $x^2 + xy + y^2 = 0$.

Some experience in algebra can be helpful, as we know that if we multiply both sides of the equation by $x - y$, we get $(x^2 + xy + y^2)$ $(x-y) = x^3 - y^3 = 0$. Therefore, $x^3 = y^3$, so that we then have $\frac{x^3}{y^3} = \left(\frac{x}{y}\right)^3 = 1$.

Algebraic Novelty 59

Problem: if $x + y = 1$, and $x^2 + y^2 = 2$, we seek $x^4 + y^4$.

Solution: One way to generate the sum of squares is to do the following:

$$(x+y)^2 = x^2 + y^2 + xy, \text{ then } 1^2 = 2 + 2xy, \text{ so that } xy = -\frac{1}{2}.$$

We are now ready to pursue $x^4 + y^4$. We can generate these fourth powers by squaring the second power expression:

$$\left(x^2 + y^2\right)^2 = x^4 + 2x^2 y^2 + y^4, \text{ which is } 2^2 = x^4 + y^4 + 2(xy)^2$$

$$= x^4 + y^4 + 2\left(-\frac{1}{2}\right)^2.$$

Thus, $x^4 + y^4 = \frac{7}{2}$.

Algebraic Novelty 60

Problem: Find all solutions of the following equation:

$$\sqrt{x+10} - \frac{6}{\sqrt{x+10}} = 5.$$

Solution: To eliminate the denominator, multiply both sides of the equation by $\sqrt{x+10}$ to get the following equation:

$$(\sqrt{x+10})(\sqrt{x+10}) - \left(\frac{6}{(\sqrt{x+10})}\right)(\sqrt{x+10}) = 5(\sqrt{x+10}),$$

so that $x + 10 - 6 = 5\left(\sqrt{x+10}\right) = x + 4.$

We now square both sides of this equation to get

$$\left(\frac{x+4}{5}\right)^2 = \left(\sqrt{x+10}\right)^2 = x + 10.$$

We then have

$$x^2 + 8x + 16 = 25(x + 10) = 25x + 250.$$

In simplified form we get the quadratic equation $x^2 - 17x - 234 = (x - 26)(x + 9) = 0$. Thus, the roots of this equation are $x = 26$ and $x = -9$.

Algebraic Novelty 61

Problem: Find the four roots of the equation $x^4 + 16x - 12 = 0$.

Solution: Using some algebraic experience, we can factor the original equation as $x^4 + 16x - 12 = (x^2 + 2^2) - 4(x - 2)^2 = (x^2 + 2x + 6) = 0$.
Since the previous product is equal to 0, each of the factors can be 0. Therefore, using the quadratic formula we solve each of the two quadratic equations to get $x^2 + 2x - 2 = 0$, so that $x = -1 \pm \sqrt{3}$, $x^2 - 2x + 6 = 0$, and then $x = -1 \pm i\sqrt{5}$.

Algebraic Novelty 62

Problem: Find the value of y for which the following pair of equations have a real common solution: $x^2 + y^2 = 16$, and $x^2 - 3y = -12$.

Solution: We will begin by substituting $x^2 = 3y - 12$ into the first equation, so that $3y - 12 + y^2 - 16 = 0$. Then we get $y^2 + 3y - 28 = (y - 4)(y + 7) = 0$. This gives us two pairs of solutions: $y = 4$, $x = 0$, and $y = -7$, $x = \sqrt{-33} = i\sqrt{33}$. The latter pair does not yield real numbers, and therefore, the former pair is the correct value for y, namely, $y = 4$.

Algebraic Novelty 63

Problem: There are two positive numbers that may be inserted between 3 and 9 so that the first 3 numbers are in a geometric progression, and the last 3 numbers are in an arithmetic progression. What is the sum of these two positive numbers?

Solution: Let these sought after positive numbers be denoted by x and y, so that the first three numbers are $3, x, y$, and the last three numbers are $x, y, 9$. Since the first three numbers are in a geometric progression, we have $\frac{x}{3} = \frac{y}{x}$, or $y = \frac{x^2}{3}$, and since the last three numbers are in an arithmetic progression, we have $y - x = 9 - y$, or $y = \frac{x+9}{2}$. Now equating the two values of y, we get $\frac{x^2}{3} = \frac{x+9}{2}$, so then $2x^2 - 3x - 27 = (2x - 9)(x + 3) = 0$. The problem required that we find positive values for x and y, therefore, $x = \frac{9}{2}$, and $y = \frac{27}{4}$. The sum of $x + y = \frac{45}{4}$.

Algebraic Novelty 64

Problem: If the following three numbers are in a geometric progression, find the fourth member of the progression: x, $2x + 2$, $3x + 3$, [?].

Solution: The common ratio in this progression is: $\frac{2x+2}{2} = \frac{3x+3}{2x+2} = \frac{3}{2}$, where $x \neq 0$, and $x \neq -1$.

Therefore, $2x + 2 = \frac{3x}{2}$, and $x = -4$, and thus, the fourth term is $(-4)\left(\frac{3}{2}\right)^3 = -13\frac{1}{3}$.

Algebraic Novelty 65

Problem: What are the integer solutions of the equation $2^{2x} - 3^{2y} = 55$?

Solution: We have $2^{2x} - 3^{2y} = (2^x + 3^y)(2^x - 3^y) = 55 = 11 \cdot 5$. Therefore, $2^x + 3^y = 11$, and $2^x - 3^y = 5$. Then by adding these two equations $2 \cdot 2^x = 16$, so that $x = 3$, and $y = 1$.

Meanwhile, another option would be that $2^x + 3^y = 55$, and $2^x - 3^y = 1$, but that does not generate integer solutions.

Algebraic Novelty 66

Problem: Find integer values of a and b which satisfy the following equation:

$$\sqrt{10+\sqrt{84}} = \sqrt{a} + \sqrt{b}.$$

Solution: Square both sides of the equation to get $10+\sqrt{84} = a+b+2\sqrt{ab}$, which is also $10+2\sqrt{21} = a+b+2\sqrt{ab}$. Therefore, $a + b = 10$, and $ab = 21$. Solving these equations by substituting $b = 10 - a$ into $ab = 21$, we get $a^2 - 10a + 21 = 0$, so that $a = 3$, and $b = 7$, or $a = 7$, and $b = 3$.

Algebraic Novelty 67

Problem: Find the value of

$$\frac{(x+b)(x+c)}{(a-b)(a-c)} + \frac{(x+c)(x+a)}{(b-c)(b-a)} + \frac{(x+b)(x+a)}{(c-a)(c-b)} = f(x).$$

Solution: This is best solved by inspecting the unusual aspects of these fractions. For example, when $x = -a$, we get

$$\frac{(-a+b)(-a+c)}{(a-b)(a-c)} + 0 + 0 = 1.$$

When $x = -b$, then

$$0 + \frac{(-b+c)(-b+a)}{(b-c)(b-a)} + 0 = 1.$$

And when $x = -c$, we have

$$0 + 0 + \frac{(-c+b)(-c+a)}{(c-a)(c-b)} = 1.$$

Since the two-dimensional function $f(x) = 1$ for the 3 values of x, it therefore must equal 1 for all values of x.

Algebraic Novelty 68

Problem: Find all real values of x such that

$$\frac{x^3 - x^2 - x + 1}{x^3 - x^2 + x - 1} = 0.$$

Solution: By factoring both numerator and denominator, we get the following:

$$\frac{x^3 - x^2 - x + 1}{x^3 - x^2 + x - 1} = \frac{x^2(x-1)-(x-1)}{x^2(x-1)+(x-1)} = \frac{(x^2-1)-(x-1)}{(x^2-1)+(x-1)} = \frac{(x-1)^2(x-1)}{(x^2-1)(x-1)}.$$

In order to make the numerator equal to 0, and not the denominator equal to 0, we find that $x = -1$. Notice that if $x = 1$, then although the numerator will be equal to 0, the denominator will be equal to 0 as well, which is unacceptable here.

Algebraic Novelty 69

Problem: If the reciprocal of $x + 1$ is $x - 1$, find the value of x.

Solution: Symbolically, we write this as $\frac{1}{x+1} = x - 1$, which yields $(x+1)(x-1) = 1$, or $x^2 - 1 = 1$. Therefore, $x^2 = 2$, and $x = \pm\sqrt{2}$.

Algebraic Novelty 70

Problem: Find the difference between the larger and smaller roots of the following equation:

$$x^2 - px + \frac{p^2 - 1}{4} = 0.$$

Solution: Using the famous quadratic formula, we get:

$$\frac{p \pm \sqrt{p^2 - 4\left(\frac{p^2-1}{4}\right)}}{2} = \frac{p \pm \sqrt{p^2 - p^2 + 1}}{2} = \frac{p \pm 1}{2}.$$

Therefore, the sought after difference is

$$\frac{p+1}{2} - \frac{p-1}{2} = 1.$$

Algebraic Novelty 71

Problem: For an arithmetic series the sum of the first 50 terms is 200, and the sum of the next 50 terms is 2,700. What is the first term of the series?

Solution: The formula for the sum S of an arithmetic series of n terms, where the first term is a, the last term is l, and the common difference is d, is $S = \frac{n}{2}(a+l)$. For the given situation, the first term is a and the last term is $a + 49d$. Now applying the formula to the series' first 50 terms, we get $200 = \frac{50}{2}(a+a+49d) = 25(2a+49d)$, or $8 = 2a+49d$.

Similarly, for the next 50 terms, the sum is 2,700, so that

$$2,700 = \frac{50}{2}\left[(a+50d) + (a+99d)\right] = 25a + 1,250d + 25a + 2,475d$$

$$= 50a + 3,725d, \text{ or } 54 = a + 74.5d.$$

Solving these two equations simultaneously yields $a = -20.5$, which is the first term of the required series.

Algebraic Novelty 72

Problem: Find the smallest value of

$$\frac{4x^2 + 8x + 13}{6(1+x)},$$

when $x \geq 0$.

Solution: We seek the smallest value of

$$y = \frac{4x^2 + 8x + 13}{6(1+x)} = \frac{4(x^2 + 2x + 1) + 9}{6(1+x)} = \frac{4(x+1)^2 + 9}{6(1+x)},$$

where $x \geq 0$. Let $x + 1 = z$; then the expression whose minimum we must determine may be written as $y = \frac{4x^2 + 9}{6x}$, and since $x \geq 0$, we have $x \geq 1$. Multiply both sides by $6x$ to get $4x^2 + 9 = 6zy$, or $4x^2 + 9 - 6zy = 0$. Now write z in terms of y:

$$z = \frac{6y}{8} \pm \frac{1}{8}\left(36y^2 - 144\right)^{\frac{1}{2}} = \frac{3y}{4} \pm \frac{3}{4}\left(y^2 - 4\right)^{\frac{1}{2}}.$$

Since z is real, $y^2 \geq 4$, so $y \geq 2$ or $y \leq -2$. However, the condition $x \geq 0$ implies that the expression for y is never negative. Therefore, the smallest value y may have is 2. When $y = 2$, $z = \frac{6 \cdot 2}{8} = \frac{3}{2}$, so that $x = \frac{1}{2}$.

Algebraic Novelty 73

Problem: Find the sum of the infinite series whose nth term is $\frac{7^{n-1}}{10^n}$.

Solution: The first term is $a_1 = \frac{1}{10}$, then $a_2 = \frac{7}{100}$. This would indicate that the common factor $r = \frac{7}{10}$. Thus, the sum of the infinite series is

$$S_\infty = \frac{a}{1-r} = \frac{\frac{1}{10}}{\frac{3}{10}} = \frac{1}{3}.$$

Algebraic Novelty 74

Problem: What is the number of solutions in positive integers of $2x + 3y = 763$?

Solution: Solving $2x + 3y = 763$ for x gives us $x = \frac{763-3y}{2}$. Since x is a positive integer, $763 - 3y$ must be a positive even number, so that y must be a positive integer, such that $3y \le 763$. There are 254 multiples of 3 less than 763, half of which are even multiples and half are odd multiples. Therefore, there are 127 possible solutions to the given equation under the stated conditions.

Algebraic Novelty 75

Problem: Find what the expression $\frac{P+Q}{P-Q} - \frac{P-Q}{P+Q}$ is equivalent to, when $P = x + y$, and $Q = x - y$.

Solution: We can immediately see that $P + Q = 2x$, and $P - Q = 2y$. Therefore,

$$\frac{P+Q}{P-Q} - \frac{P-Q}{P+Q} = \frac{2x}{2y} - \frac{2y}{2x} = \frac{x^2 - y^2}{xy}.$$

Chapter 4

Geometric Novelties

Most high school students in the United States take one year of geometry class. They are exposed to a logical system involving triangles, quadrilaterals, other polygons, and circles, and learn these shapes' unusual properties and interactions with one another. However, there are many novel relationships in plane geometry that are typically not taught during the one-year geometry course. This chapter explores a plethora of somewhat unusual geometric relationships. It will hopefully enhance the reader's appreciation for geometry as a truly beautiful aspect of mathematics.

Geometric Novelty 1

Problem: Three rectangles, $ABCD$, $CGFE$, and $GHKJ$, are placed together, as shown in Figure 4-1, and a straight line is drawn connecting point B to point K. The dimensions of rectangle $ABCD$ are $AB = 1$ and $AD = 2$. The dimensions of rectangle $CGFE$ are $GF = 2$ and $EF = 4$. The dimensions of rectangle $GHKJ$ are $KH = 4$ and $KJ = 8$. What is the area of the shaded region?

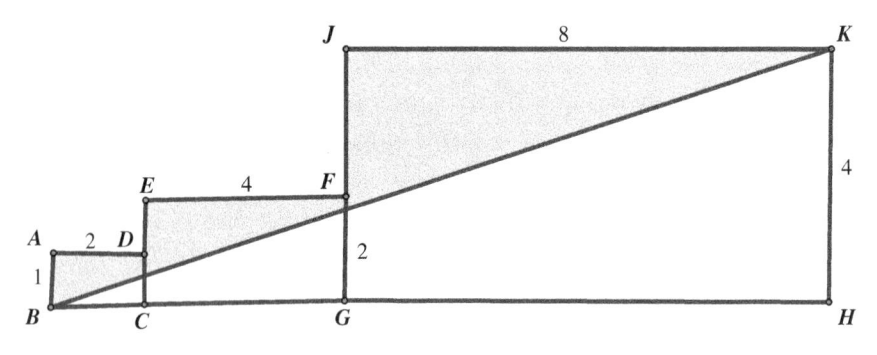

Figure 4-1

Solution: The figure whose area we seek is rather oddly shaped. It is natural to first try to find the area of each of the shaded regions. However, this would prove impossible since we do not know where *BK* intersects the rectangles' sides. Let's approach this problem from a different point of view. We can find the area of the three rectangles and then subtract the area of the unshaded region, which is simply a right triangle whose legs are 14 and 4. Rectangle *ABCD* has an area of $2 \cdot 1 = 2$, rectangle *CGFE* has an area of $2 \cdot 4 = 8$, and rectangle *GHKJ* has an area $4 \cdot 8 = 32$. Therefore, the sum of the areas of the three rectangles is 42. The area of right triangle $BHK = \frac{1}{2} \cdot 14 \cdot 4 = 28$. Thus, the area of the shaded region is equal to the sum of the areas of the three rectangles minus the area of the right triangle, or $42 - 28 = 14$. Adopting a different point of view can often take a seemingly complex problem from the realm of the time-consuming and quite difficult, into the realm of a relatively simple problem.

Geometric Novelty 2

Problem: If a solid ball 3″ in diameter weighs 7 pounds, how many pounds are there in the weight of a ball of the same material 6″ in diameter?

Solution: We may assume the following relationships when comparing similar figures:

- (ratio of corresponding linear parts)^2 = ratio of corresponding areas,

- (ratio of corresponding linear parts)^3 = ratio of corresponding volumes.

Linear parts may be radii, circumferences, diameters, sides, edges, diagonals, perimeters, and so on. We employ the second relationship for this problem, since we are comparing volumes with linear diameters. Keep in mind that the weight of a solid, such as a ball, is proportional to its volume. Therefore, since the ratio of the corresponding linear parts (diameters) is $\frac{3}{6} = \frac{1}{2}$, the ratio of the volumes is $\left(\frac{1}{2}\right)^3$ or $\frac{1}{8}$. Hence, the weight of the second ball is 8 times the weight of the first ball, or 56 pounds.

Geometric Novelty 3

Problem: In Figure 4-2, quadrilaterals *ABCD* and *BDFE* are parallelograms. What is the ratio of their areas?

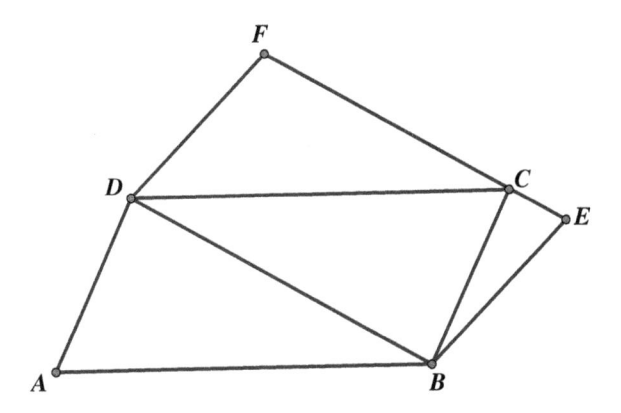

Figure 4-2

Solution: An immediate reaction might be to examine each of the parallelograms and try to find their individual areas. Instead, let's examine the problem from a different point of view. In parallelogram *ABCD*, triangle $BCD = \frac{1}{2}$ of the area of *ABCD*, since *BD* is a diagonal. In parallelogram *BDEF*, triangle $BCD = \frac{1}{2}$ the area of *DBFE*, since they have a common base *BD* and the same altitude, which is the perpendicular from *C* to *BD*. Thus, since triangle *BCD* has half the area of the parallelograms, *ABCD* and *DBFE* must therefore be equal in area.

Geometric Novelty 4

Problem: Find the difference between the areas of the two right triangles *ABC* and *BCD*, shown in Figure 4-3, where $BC = 18$, $BD = 30$, and $AD = 9$.

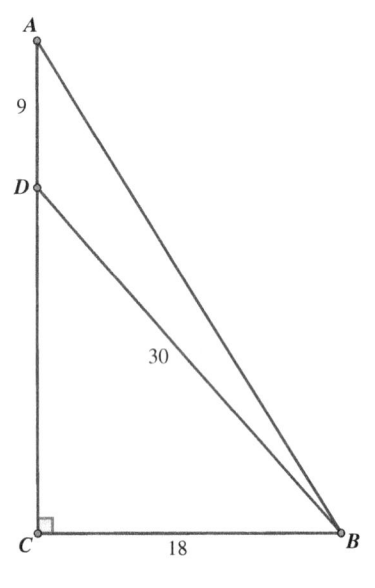

Figure 4-3

Solution: The typical approach to solving this problem would be to find the length of *DC* by applying the Pythagorean theorem, where $DC^2 = 30^2 - 18^2 = 24^2$, so that $DC = 24$. Then we find the area of triangle $ABC = \frac{1}{2} \cdot (24+9) \cdot (18) = 297$, and the area of triangle $DCB = \frac{1}{2}(24)(18) = 216$. Thus, the difference between the areas of these two triangles is $297 - 216 = 81$.

However, there is a much simpler way of approaching the problem. We revisit the question and find that it asks for "the difference between the areas." This, geometrically speaking, refers to triangle ADB, whose area can be easily obtained since its base length is $AD = 9$, and its altitude $BC = 18$. Once again, we get the area of triangle

$$ADB = \frac{1}{2} \cdot 9 \cdot 18 = 81.$$

Geometric Novelty 5

Problem: In Figure 4-4, Triangle ABC is equilateral and polygon $DEFGHJ$ is a regular hexagon. If $FG = \frac{1}{3}AB$, what is the ratio of the area of the hexagon to triangle ABC?

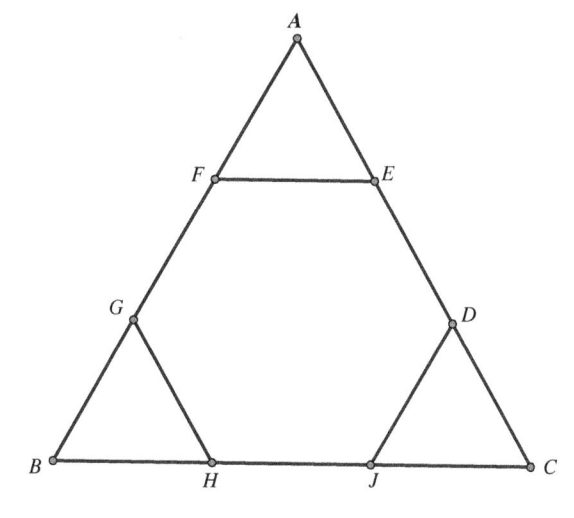

Figure 4-4

Solution: The traditional way of responding to this question is to find the area of hexagon $DEFGHJ$ by letting the length of side $FG = 1$, and then the sides of equilateral triangle ABC will have length 3. However, since we are looking for the relative areas of these two figures it would be much easier to divide the hexagon into the six equilateral triangles as shown in Figure 4-5.

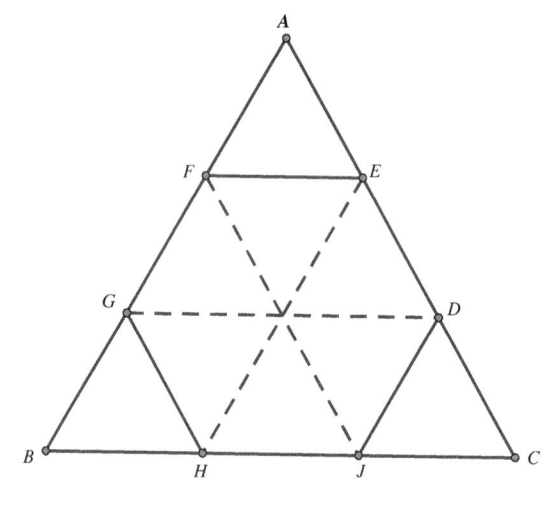

Figure 4-5

We find that of the 9 equilateral triangles comprising triangle *ABC*, 6 of these comprise hexagon *DEFGHJ*. Therefore, the ratio of the area of the hexagon to the equilateral triangle *ABC* is 6:9 or 2:3.

Geometric Novelty 6

Problem: In Figure 4-6, we have rectangle *ABCD* with $AB = 12$, $BC = 9$, and $EB = BF = 3$. Find the area of triangle *DEF*.

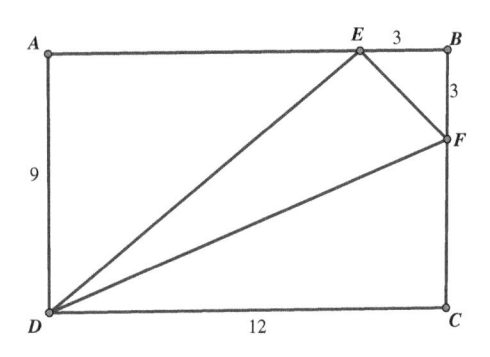

Figure 4-6

Solution: A natural tendency in seeking the area of a triangle is to find its dimensions. The dimensions of *DEF* can be obtained by applying the Pythagorean theorem to the three triangles whose hypotenuses form the three sides of triangle *DEF*. We would then need to apply Heron's formula[1] for the area of the triangle given only the lengths of the sides.

However, a much more efficient procedure would be to find the sum of the areas of the three right triangles, *ADE*, *DCF*, and *EBF*, and subtract that from the area of the rectangle. The area of triangle $ADE = \frac{1}{2} \, 9 \cdot 9 = 40.5$, the area of triangle $DCF = \frac{1}{2} \, 12 \cdot 6 = 36$, and the area of triangle $EBF = \frac{1}{2} \, 3 \cdot 3 = 4.5$, so that the sum of the areas of these three triangles is 81. The area of rectangle $ABCD = 9 \cdot 12 = 108$. Thus, the area of triangle $DEF = 108 - 81 = 27$.

Geometric Novelty 7

Problem: Square *ABCD* in Figure 4-7 has sides length 12 and the trisection points of the sides are *E, F, G,* and *H*. Find the area of quadrilateral *PQRS*.

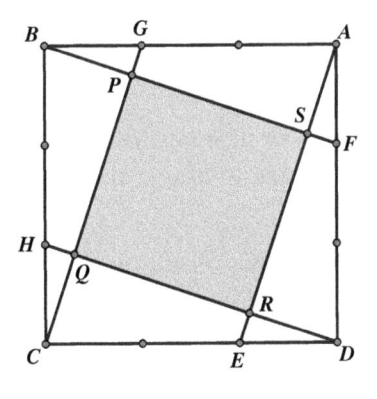

Figure 4-7

Solution: The following four right triangles are congruent: Δ*ASF*, Δ*BPG*, Δ*CQH*, and Δ*DRE*. They are similar, respectively, to the following other right triangles: Δ*ADE*, Δ*BGC*, Δ*CHD*, and Δ*DAE*, with a ratio of 1:3. Therefore, as shown in Figure 4-8, let $HQ = x$, and $CQ = 3x$. We also have $RD = 3x$. Because of the similarity of the right triangles, we know

---

[1] Heron's Formula: To find the area of a triangle with side lengths a, b, c, where s is the semi-perimeter, we use $\sqrt{s(s-a)(s-b)(s-c)}$.

that $BG = \frac{1}{3}BC$, so that $CQ = \frac{1}{3}QD$, thus, $QR = 6x$. Applying the Pythagorean theorem to triangle CQD, we find that $CD = 3x\sqrt{10} = 12$, and then $x = \frac{2\sqrt{10}}{5}$. Therefore, the area of square $PQRS$ is

$$QR^2 = (6x)^2 = \left(6 \cdot \frac{2\sqrt{10}}{5} \right)^2 = 36 \cdot \frac{4 \cdot 10}{25} = \frac{288}{5}.$$

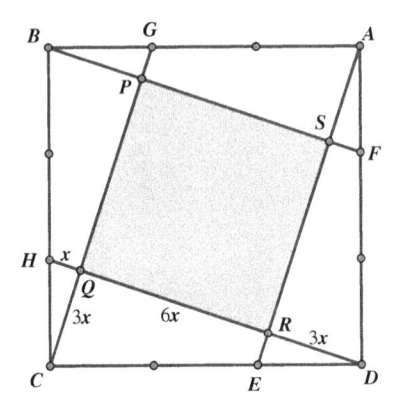

Figure 4-8

Geometric Novelty 8

Problem: Two concentric circles, shown in Figure 4-9, are 12 units apart. What is the difference between the circumference lengths of the two circles?

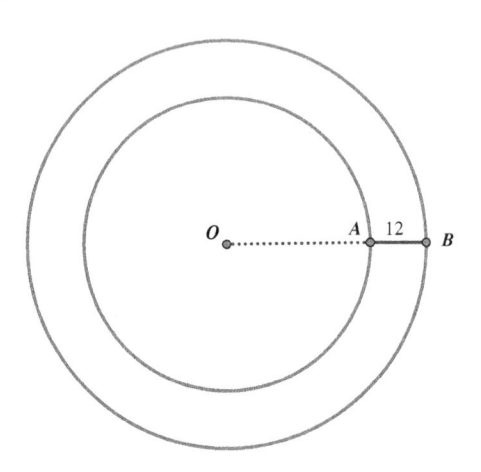

Figure 4-9

Solution: The traditional solution to this question is to find the circumference of each of the two circles by letting the radius $AO = r$, so that the circumference of the larger semicircle is $2\pi(r + 12)$, and the circumference of the smaller semicircle is $2\pi r$. Therefore, the difference between the two circumference lengths is simply $2\pi(r + 12) - 2\pi r = 24\pi$.

However, a much simpler way to respond to this question is as follows. Since the relative sizes of the circles is not known, we can assume, without loss of generality, that the smaller circle becomes minuscule or in effect simply a point. In that case, the distance between the two concentric circles is actually the radius of the larger circle, namely, 12. Therefore, the difference between the circumference lengths is $2\pi(12) - 0 = 24\pi$.

Geometric Novelty 9

Problem: Given a randomly drawn pentagram, as shown in Figure 4-10, what is the sum of the angles at the vertices A, B, C, D, and E?

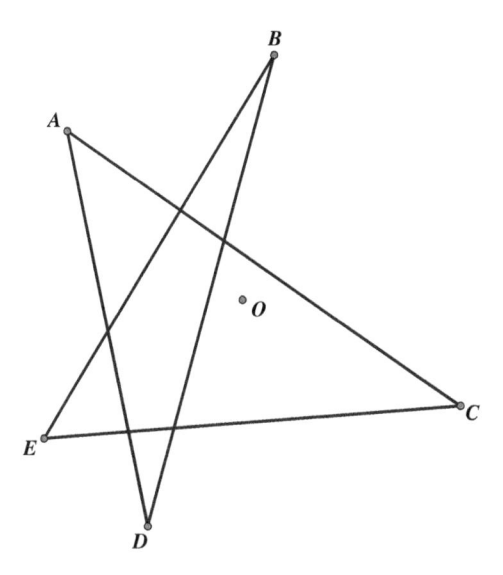

Figure 4-10

Solution: There are various ways to approach this problem. One can assume that this randomly drawn pentagram could just as easily have been drawn with equal angles. In that case, the inner figure would be a regular pentagon, as shown in Figure 4-11, whose angle sum is 540° and whose individual angles measure 108°. We then note that an adjacent isosceles triangle, for example isosceles triangle *BFG*, containing one of the vertices of the pentagram has base angles of 72°, leaving 36° for a vertex angle. Therefore, the sum of the vertex angles is $5 \cdot 36° = 180°$.

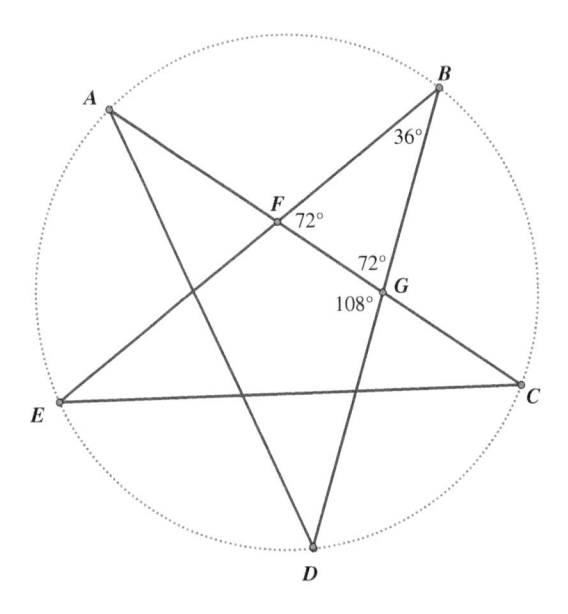

Figure 4-11

A more efficient method of solution would be to consider a randomly drawn pentagram that is circumscribed by a circle, as shown in Figure 4-12. We notice that each of the vertex angles is measured by one-half of the intercepted arc. For example, $\angle A = \frac{1}{2}\overset{\frown}{DC}$. When this is repeated for each of the angles, we find that the sum of the vertex angles is one half the sum of the arcs, or simply put:

$$\frac{1}{2}360° = 180°.$$

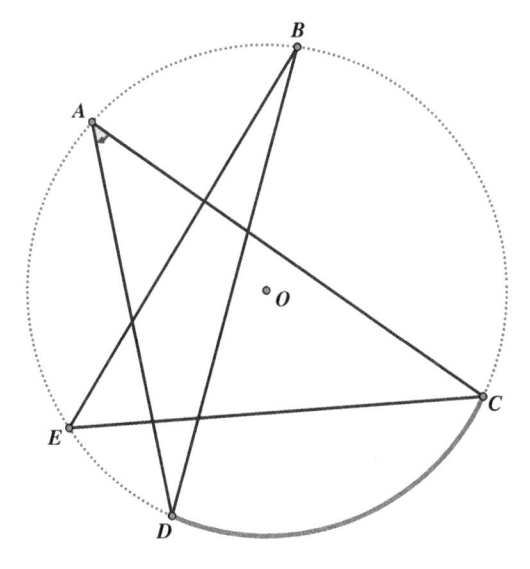

Figure 4-12

Geometric Novelty 10

Problem: In Figure 4-13, the two circles and the three small semicircles are tangent to each other and to the large semicircle at various points of contact. The two smaller circles and the three semicircles each have a radius of one unit. Find the area of the shaded region shown in Figure 4-13.

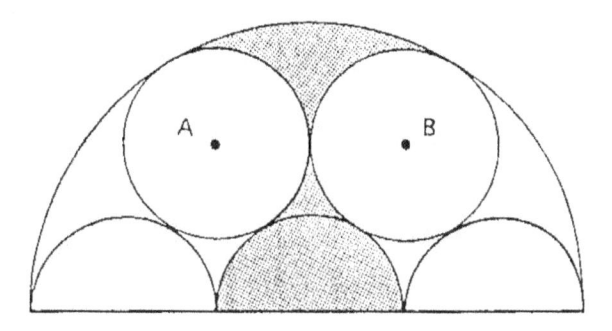

Figure 4-13

Solution: Consider the same diagram, as shown in Figure 4-14, where we can consider the two circles and three semicircles removed. Now considering the three dark shaded regions are actually three congruent regions, one of which is the required region sought in the original problem. Therefore, to get the dark shaded region in Figure 4-14, we simply find the area of the large semicircle, whose radius is 6 and, therefore, $\frac{1}{2}\pi 3^2 = \frac{9}{2}\pi$. The area of the 2 circles and 3 semicircles is

$$2\pi 1^2 + 3\left(\frac{1}{2}\pi 1^2\right) = \frac{7}{2}\pi.$$

Thus, the area of the dark shaded region is

$$\frac{9}{2}\pi - \frac{7}{2}\pi = \pi.$$

Therefore, the required shaded region in Figure 4-13 is one-third of the dark shaded region shown in Figure 4-14, which is $\frac{\pi}{3}$, plus the area of one semicircle:

$$\frac{\pi}{3} + \frac{\pi}{2} = \frac{5}{6}\pi.$$

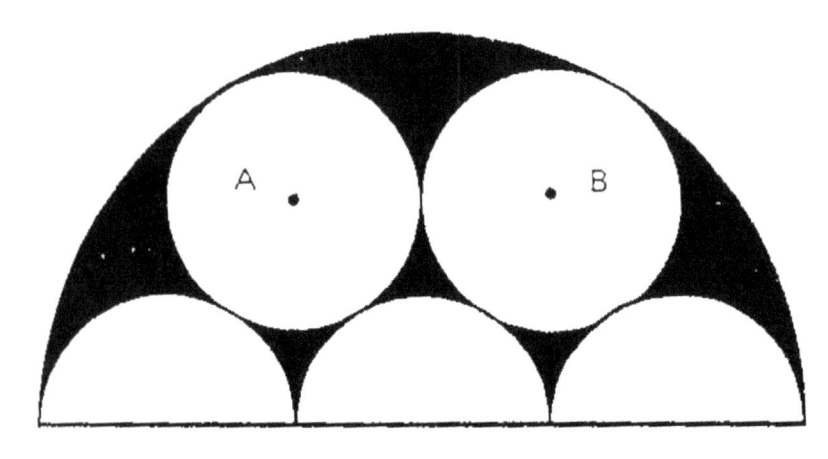

Figure 4-14

Geometric Novelty 11

Problem: Figure 4-15 depicts square $ABCD$. Arc AFC is part of a circle centered at D, and arc AEC is part of a circle centered at B. The area of square $ABCD$ is 4 square units. Find the area between the two arcs (the darker shaded region).

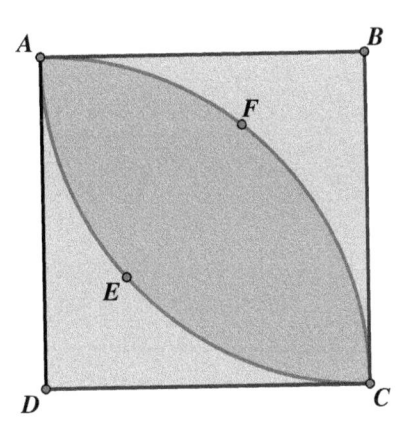

Figure 4-15

Solution: A common method to solve this problem would be to find the area of sector

$$ABCE \left(\frac{1}{4} \cdot \pi \cdot 2^2 = \pi \right),$$

subtract from that sector the area of triangle ABC (2) to get $(\pi - 2)$, and then double that area to provide the required darker shaded area, $2\pi - 4$.

However, perhaps a more sophisticated way to solve this problem would be to take the area of sector $ABCE$, which is a quarter of a circle with radius 2, which is π, and then find the area of sector $ADCF$, which is also π. By adding these two areas to get 2π and subtracting the area of the square, we get $2\pi - 4$, which is the area of the darker shaded portion.

Geometric Novelty 12

Problem: In Figure 4-16, trapezoid *ABCD* has *AB||CD*. Also, $\angle D = 60°$, and $\angle C = 90°$. Find the length of *AD*, if *AD = CD*, and *AB = 14*.

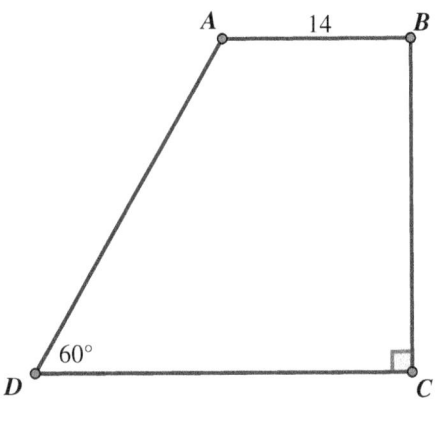

Figure 4-16

Solution: In Figure 4-17, by drawing line *AC*, we then recognize that since $\angle D = 60°$ and *AD = CD*, we have equilateral triangle *ADC*. By drawing altitude *AE*, rectangle *ABCE* results. Therefore, *DE = CE = 14*, so that *AD = DC = 28*.

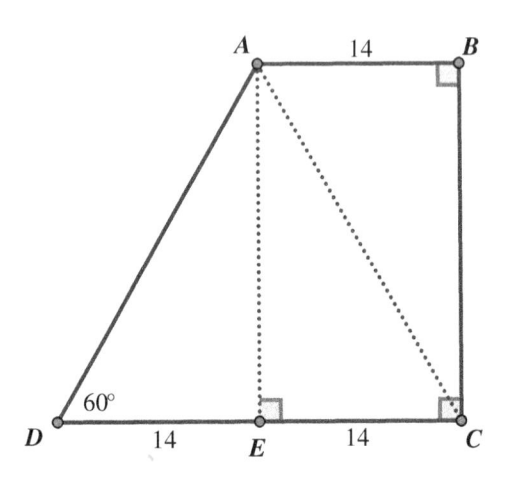

Figure 4-17

Geometric Novelty 13

Problem: In Figure 4-18, point *D* is in the interior of triangle *ABC* with the various angle sizes marked with *x*, *y*, *w*, and *z*. The challenge here is to express *x* in terms of *y*, *w*, and *z*.

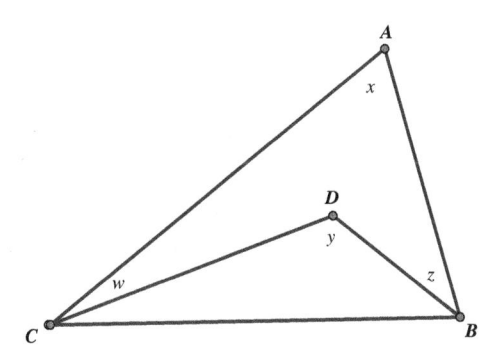

Figure 4-18

Solution: Since $x = \angle A = 180° - (w + \angle DCB) - (z + \angle DBC) = (180° - \angle DCB - \angle DBC) - w - z = y - w - z$. Since $y = 180° - \angle DCB - \angle DBC$, we have $x = y - w - z$.

Geometric Novelty 14

Problem: In Figure 4-19, the side length of square *ABCD* is 2 units and points *E* and *F* are the midpoint of sides *AD* and *AB*, respectively. The measure of angle *ECF* is represented by θ. What is $\sin \theta$ equal to?

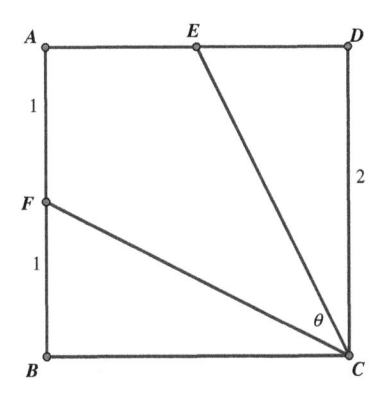

Figure 4-19

Solution: We can express the area of triangle *ECF* in two ways. (See Figure 4-20.) The first method is to use the formula

$$\text{area}\triangle ECF = \frac{1}{2}(EC)(FC)\sin\theta.$$

We apply the Pythagorean theorem applied to triangles *FBC* and *EDC* to find that that $FC = EC = \sqrt{5}$. Therefore, the

$$\text{area}\triangle ECF = \frac{1}{2}(\sqrt{5})^2 \sin\theta = \frac{5}{2}\sin\theta.$$

However, the area of triangle *EFC* can also be expressed by subtracting the area of the three right triangles from the area of the square as follows:

$$\text{area}\triangle ECF = \text{area}ABCD - \text{area}\triangle EAF - 2\text{area}\triangle FBC = 4 - \frac{1}{2} - 2(1) = \frac{3}{2}.$$

Equating these two results of both methods gives us

$$\frac{5}{2}\sin\theta = \frac{3}{2},$$

so that

$$\sin\theta = \frac{3}{5}.$$

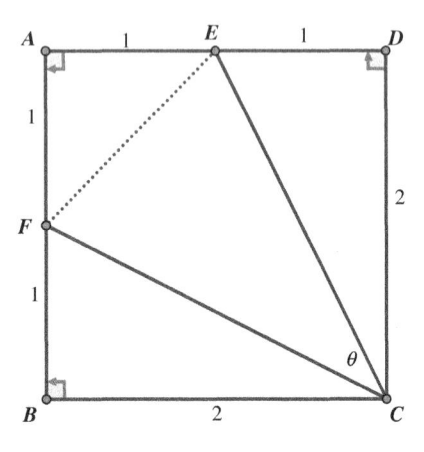

Figure 4-20

Geometric Novelty 15

Problem: In Figure 4-21, square *DEBF*, whose area is 441, is inscribed in isosceles right triangle *ABC*, where *AB* = *BC*. In Figure 4-22, square *JKLM* is inscribed in isosceles right triangle *ABC*, which is congruent to the isosceles right triangle *ABC* in Figure 4-21. Find the area of square *JKLM*.

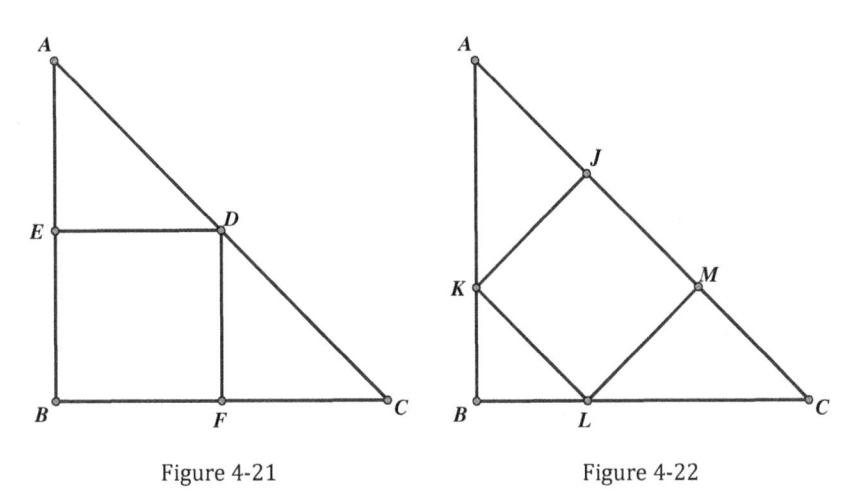

Figure 4-21   Figure 4-22

Solution: In Figure 4-21, since triangle *ABC* is isosceles, point *D* is the midpoint of *AC*. Since *ED* is parallel to *BC*, point *E* is the midpoint of *AB* and, therefore, *AB* = 2*BE*. Since the area of square *DEBF* = 441, its side length is $\sqrt{441} = 21$ and, thus, *AB* = 42. In Figure 4-23, we let *KL* = *x*,

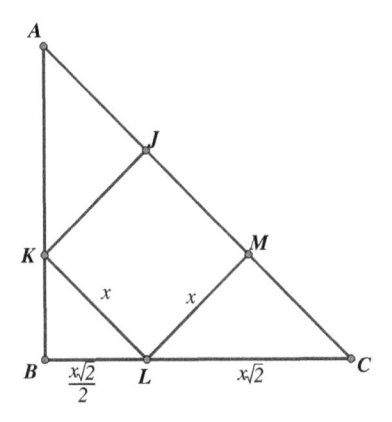

Figure 4-23

and apply the Pythagorean theorem to isosceles triangles KBL and CML to get $BL = \frac{x\sqrt{2}}{2}$ and $CL = x\sqrt{2}$. Since we have $AB = 42$, we can express

$$BC = \frac{x\sqrt{2}}{2} + x\sqrt{2} = 42,$$

and then $x = 14\sqrt{2}$.

Therefore, the area of square

$$JKLM = x^2 = \left(14\sqrt{2}\right)^2 = 392.$$

An alternate solution to this problem can be found. Because of the unusual symmetry in the given situation, we draw dashed lines, as shown in Figure 4-24, to form nine congruent isosceles right triangles. Since the area of square $DEBF = 441$, from Figure 4-21 we know that the area of isosceles right triangle $ABC = 882$. In Figure 4-24, each of the nine congruent right triangles has an area of 98, and therefore, the area of square

$$JKLM = 4 \cdot 98 = 392.$$

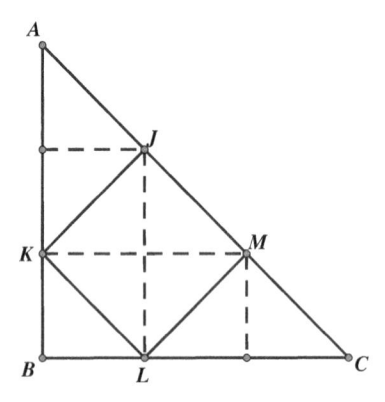

Figure 4-24

Geometric Novelty 16

Problem: In Figure 4-25, equilateral triangles ABC and DEF have parallel sides and share the same centroid, point G. The distance between the corresponding sides of the two equilateral triangles is $\frac{1}{6}$ of the altitude of triangle *ABC*. What is the ratio of the areas of the two equilateral triangles?

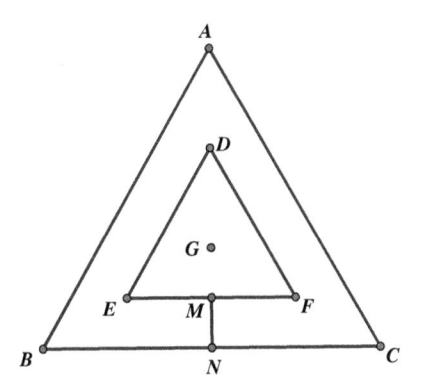

Figure 4-25

Solution: We begin by recognizing that $MN = \frac{1}{6} AN$ as shown in Figure 4-26, where *AN* is the altitude of triangle *ABC*. Since the centroid of an equilateral triangle trisects each altitude, we have $GN = \frac{1}{3} AN$ and also $GM = \frac{1}{6} AN$. However,

$$DM = 3GM = 3\left(\frac{1}{6} AN\right) = \frac{1}{2} AN.$$

Since the ratio of the attitudes of the two triangles *DEF* and *ABC* is $\frac{1}{2}$, the ratio of their areas is $\frac{1}{4}$.

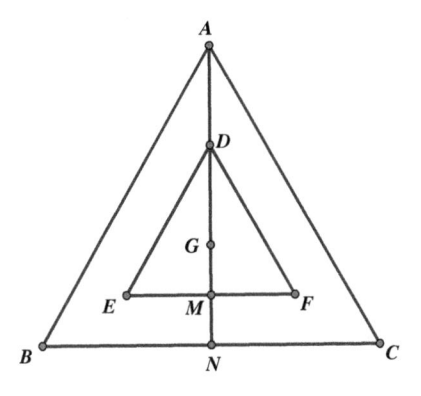

Figure 4-26

Geometric Novelty 17

Problem: In Figure 4-27, the rectangle *HIJK* has the same area as square *ABCD*, as the partitioned parts of the square have been rearranged to form the rectangle. Points *E* and *F* are the midpoints of sides *AB* and *CD* of square *ABCD*, and *AG* is perpendicular to *DE*. We seek the ratio of the length to the width of rectangle *HIJK*.

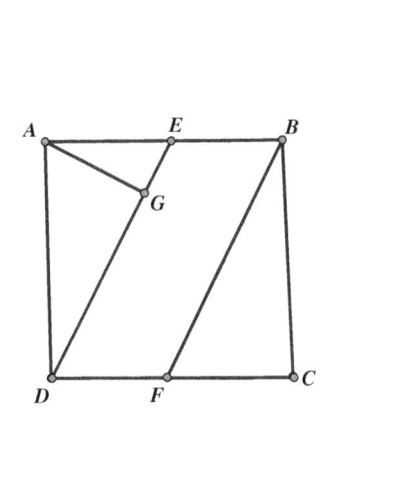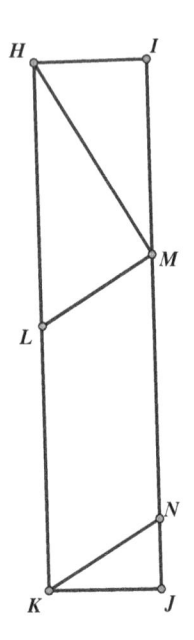

Figure 4-27

Solution: We notice that *HL* = *BF* and *KL* = *DE*. If we let the length of *AD* = *BC* = 2, we can apply the Pythagorean theorem to right triangle *BCF* to get

$$BF = \sqrt{2^2 + 1^2} = \sqrt{5}.$$

The length of the rectangle side *HK* = *HL* + *KL* = *BF* + *DE* = $2\sqrt{5}$. Since the areas of the two quadrilaterals are equal, we find that the area of rectangle

$$HIJK = 4 = KJ \cdot HK;$$

therefore,

$$KJ = \frac{4}{HK} = \frac{4}{2\sqrt{5}}.$$

Thus, the ratio of the length to the width of the rectangle is

$$\frac{2\sqrt{5}}{\dfrac{4}{2\sqrt{5}}} = \frac{\left(2\sqrt{5}\right)^2}{4} = \frac{20}{4} = 5.$$

Geometric Novelty 18

Problem: In Figure 4-28, triangle *ABC* has sides *AC* = 3, *BC* = 4, and *AB* = 5. A semicircle with center *D* on side *AC* is tangent to *AB* at point *E* and is tangent to side *BC* at point *C*. Find the radius of the semicircle.

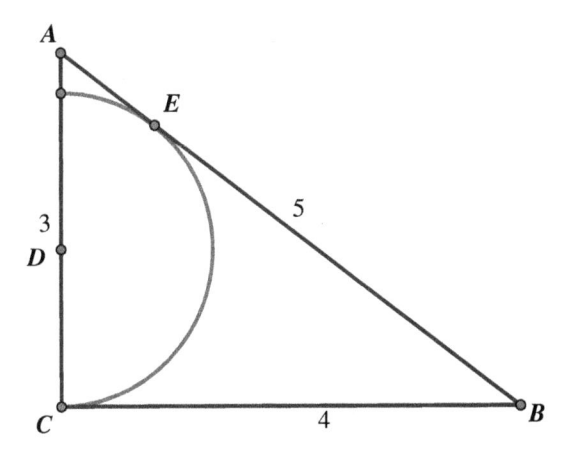

Figure 4-28

Solution: The distance from an external point to tangents to a circle are equal. Therefore, *BC* = *BE* = 4 and *AE* = 1. We let the radius of the semicircle *DC* = *DE* = *r*. We then have *AD* = 3 − *r*. Applying the Pythagorean theorem the right triangle *ADE* (since *DE* ⊥ *AB*), we get

$$r^2 + 1^2 = (3-r)^2 = 9 - 6r + r^2.$$

$$6r = 8, \text{ so that } r = \frac{4}{3}.$$

Geometric Novelty 19

Problem: In Figure 4-29, three externally tangent circles with centers D, F, and E have radii of lengths 3, 2, and 1, respectively. The three circles are tangent at points A, B, and C. Find the area of triangle ABC.

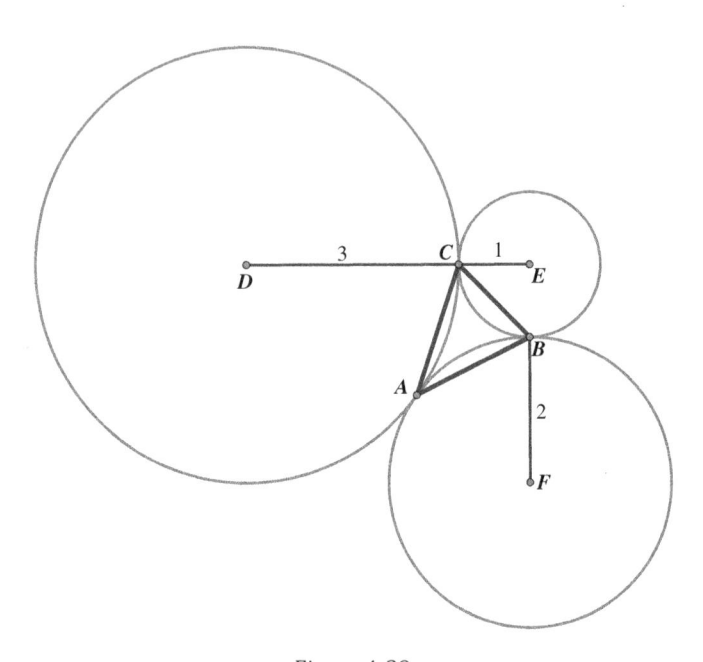

Figure 4-29

Solution: In Figure 4-30, the lines joining the centers of the three circles contain the points of tangency. As a result, we find that the problem has become simple because the sides of triangle DEF are 3, 4, and 5, thus making it is a right triangle. The area of right triangle DEF is $\frac{1}{2} \cdot 4 \cdot 3 = 6$, and from that area we subtract the areas of the three triangles $\triangle CEB$, $\triangle BFA$, and $\triangle ADC$. We find the area of $\triangle CEB$ as follows:

$$\text{area} \triangle CEB = \frac{1}{2}(1^2) = \frac{1}{2}.$$

To find the areas of the other two triangles, we use the sine formula for area. In right triangle *DEF* we first need to find

$$\sin \angle BFA = \frac{DE}{DF} = \frac{4}{5}, \text{ and } \sin \angle EDF = \frac{EF}{DF} = \frac{3}{5}.$$

Now applying the area formulas:

$$\text{area}\triangle FBA = \frac{1}{2}(AF)(BF)\sin \angle BFA = \frac{1}{2}(2)(2)\frac{4}{5} = \frac{8}{5},$$

and

$$\text{area}\triangle ADC = \frac{1}{2}(AD)(CD)\sin \angle EDF = \frac{1}{2}(3)(3)\frac{3}{5} = \frac{27}{10}.$$

We are now ready to find the area of triangle *ABC*:

$$\text{area}\triangle ABC = \text{area}\triangle DEF - \left(\text{area}\triangle CEB + \text{area}\triangle FBA + \text{area}\triangle ADC\right)$$

$$= 6 - \left(\frac{1}{2} + \frac{8}{5} + \frac{27}{10}\right) = \frac{6}{5}.$$

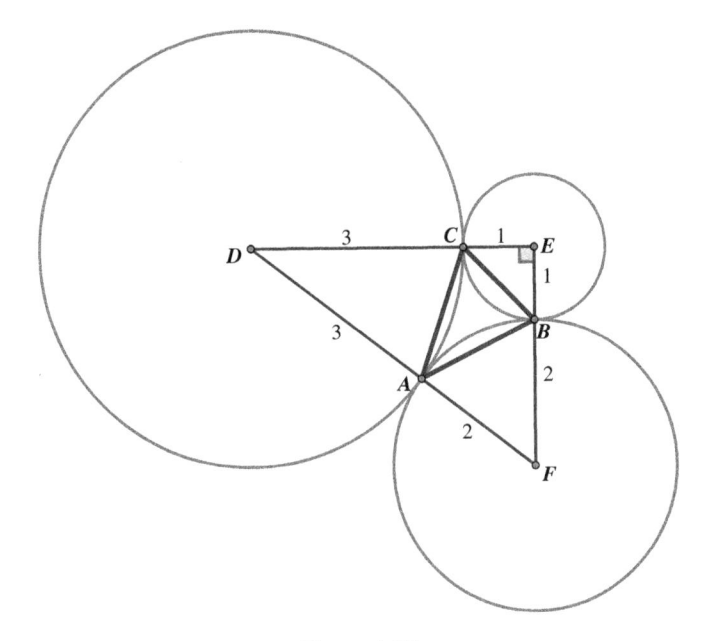

Figure 4-30

Geometric Novelty 20

Problem: In Figure 4-31, a circle is circumscribed around equilateral triangle *ABC*, and a second circle is inscribed in triangle *ABC*. What is the ratio of the areas of the two circles?

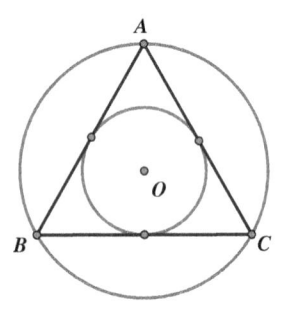

Figure 4-31

Solution: Consider the altitude *AM* of triangle *ABC*, where point *O* is the trisection point of the altitudes, as shown in Figure 4-32. We then have *AO* as the radius of the circumscribed circle and *OM* as a radius of the inscribed circle. Since the radii ratio $\frac{OM}{OA} = \frac{1}{2}$, the ratio of the areas of the two circles is $\frac{1^2}{2^2} = \frac{1}{4}$.

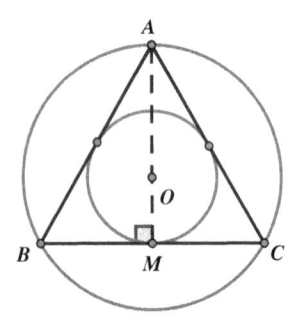

Figure 4-32

Geometric Novelty 21

Problem: In Figure 4-33, triangle *ABC* is inscribed in a circle, where *AD* is tangent to the circle and *AB* bisects angle *CBD*. If *BC* = 8, and *AB* = 6, what is the length of *BD*?

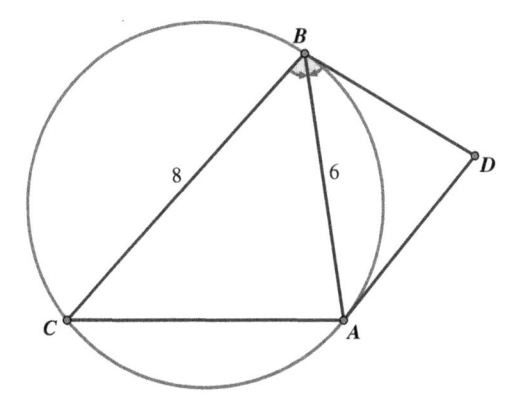

Figure 4-33

Solution: Since the angle *BAD* formed by tangent *DA* and chord *BA* is measured by one-half arc *AB*, and inscribed angle *BCA* is also measured by one-half arc *AB*, we have $\angle BAD = \angle BCA$. Because of the angle bisector *AB*, we also have $\angle ABC = \angle ABD$. Therefore, we are able to establish that $\triangle ABC \sim \triangle ADB$, so that $\frac{AB}{BC} = \frac{BD}{AB}$ and then $\frac{6}{8} = \frac{BD}{6}$. Thus, $BD = 4\frac{1}{2}$.

Geometric Novelty 22

Problem: In Figure 4-34, the exterior and interior angle bisectors at vertices *A* and *B* of triangle *ABC* meet at points *E* and *D*, respectively. If $\angle ADB = 120°$, find the measure of $\angle AEB$.

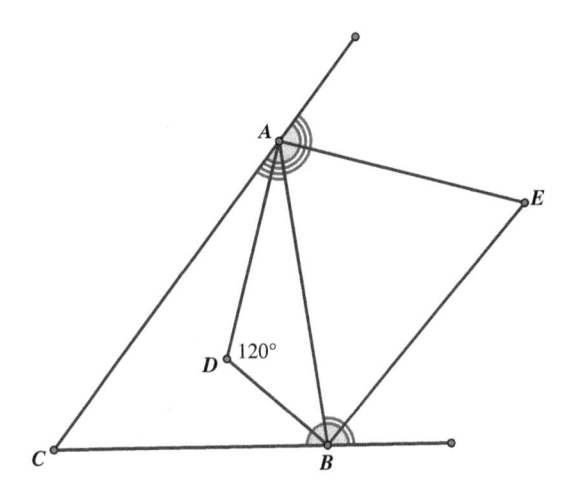

Figure 4-34

Solution: The exterior angle bisector of triangle ABC at vertex A is perpendicular to the interior angle bisector of angle CAB since angle DAE is one-half of the straight angle. Analogously, we have $DB \perp EB$. Since the opposite angles of quadrilateral $ADBE$ are right angles and, therefore, supplementary, quadrilateral $ADBE$ is cyclic. This means that angle D and angle E are also supplementary. Thus, since $\angle D = 120°$, it follows that its supplement $\angle E = 60°$.

Geometric Novelty 23

Problem: Three circles of radius length 1 are mutually tangent as shown in Figure 4-35. Find the unshaded area between the three circles.

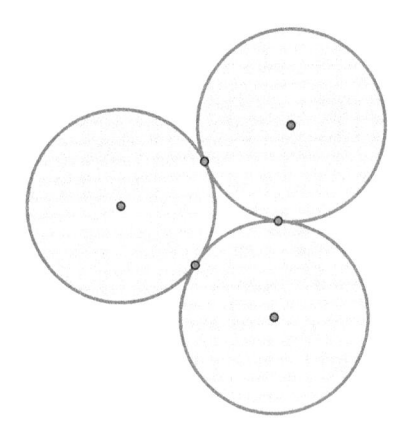

Figure 4-35

Solution: The lines joining the centers of the three congruent circles create equilateral triangle ABC with side length 2, as shown in Figure 4-36. Its area is $\frac{2^2\sqrt{3}}{4} = \sqrt{3}$. The area of one of the circles is π and the area of one the sectors (for example, EBF) is $\frac{1}{6}$ of the area of the circle, or $\frac{\pi}{6}$. Therefore, subtracting the three sectors EBF, EAD, and DCF from the area of the equilateral triangle ABC gives us

$$\sqrt{3} - 3 \cdot \frac{\pi}{6} = \sqrt{3} - \frac{\pi}{2}.$$

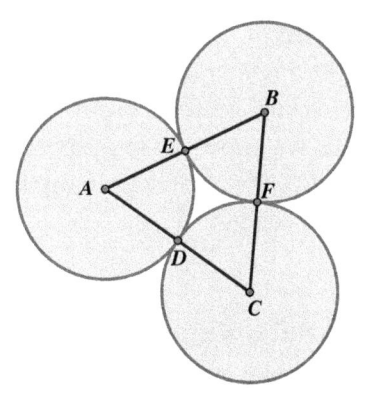

Figure 4-36

Geometric Novelty 24

Problem: Figure 4-37 shows a cube with side lengths one unit, where vertex A is opposite side square $CDEG$, and where the intersection of the diagonals of $CDEG$ is point H. Find the distance from point A to point H.

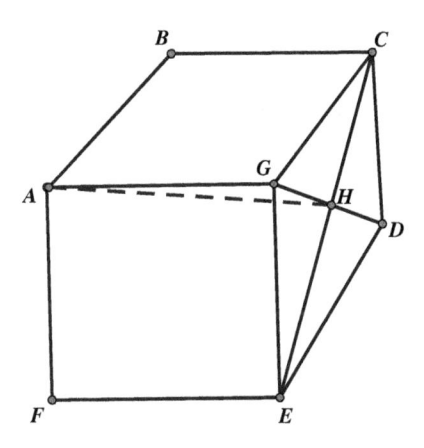

Figure 4-37

Solution: To find the length of AH, we need to immediately recognize that angle AGH is a right angle and triangle AGH is a right triangle so that we can apply the Pythagorean theorem.

In right triangle CGE, we know the length of diagonal CE is $\sqrt{2}$, and therefore, the length of $GH = \frac{\sqrt{2}}{2}$. Now applying the Pythagorean theorem to right triangle AGH, we get

$$AH^2 = AG^2 + GH^2 = 1^2 + \left(\frac{\sqrt{2}}{2}\right)^2 = \frac{3}{2}.$$

Therefore, $AH = \sqrt{\frac{3}{2}} = \frac{\sqrt{6}}{2}$.

Geometric Novelty 25

Problem: Shown in Figure 4-38, three mutually congruent circles of radius r and with centers at A, B, and C are circumscribed by a circle with radius R and tangent at points G, H, and K. What is the radius of the large circle in terms of r?

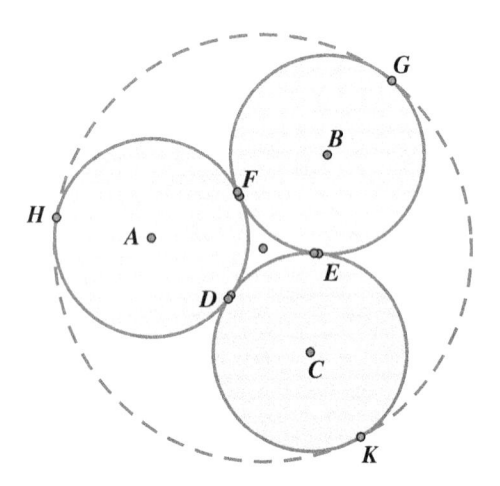

Figure 4-38

Solution: In Figure 4-39, equilateral triangle ABC has altitude BD, which contains the center point P of the large circle whose radius is R.

Point P is the trisection point of altitude BD, whose length obtained by the Pythagorean theorem is

$$BD^2 = AB^2 - AD^2 = (2r)^2 - r^2 = 3r^2, \text{ so that } BD = r\sqrt{3}.$$

Therefore,

$$PB = \frac{2}{3}BD = \frac{2r\sqrt{3}}{3}.$$

Thus,

$$R = PG = PB + BG = \frac{2r\sqrt{3}}{3} + r = \frac{3r + 2r\sqrt{3}}{3}.$$

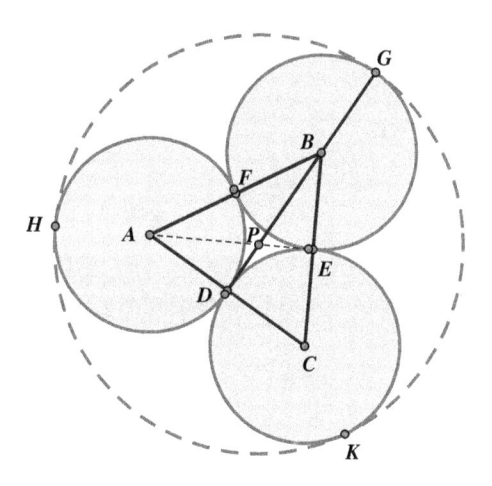

Figure 4-39

Geometric Novelty 26

Problem: The angles of one triangle have degree measures of x, y, and z. The angle measures of the second triangle are $k - x$, $k - y$, and $k - z$. Under these circumstances, what is the numerical value of k?

Solution: Clearly the sum of the angles of the first triangle $x + y + z = 180°$. The sum of the angles of the second triangle is $180° = (k - x) + (k - y) + (k - z) = 3k - (x + y + z) = 3k - 180°$, so that $3k = 360°$, and thus, $k = 120°$.

Geometric Novelty 27

Problem: Square *ABCD* has side length 1. Four quarter circles centered at each of the four vertices are inscribed in the square and create a four-sided figure *EFGH*, as shown in Figure 4-40. A circle is inscribed into this four-arced figure centered at *P*, the center of the square. Find the area of this inscribed circle.

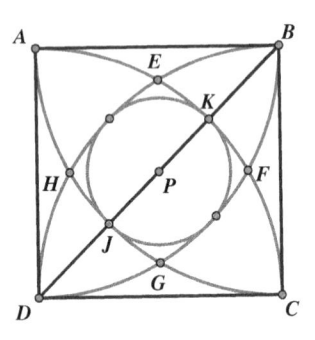

Figure 4-40

Solution: Given the length of side *AB* of square *ABCD* is 1, we therefore have $AB = BJ = 1 = DK$. We apply the Pythagorean theorem to triangle *ABD* to get $DB = \sqrt{2}$.

Consider

$$KJ = BJ + DK - BD = 1 + 1 - \sqrt{2} = 2 - \sqrt{2};$$

therefore, the radius of the inscribed circle is $\frac{2-\sqrt{2}}{2}$. Thus, the area of the inscribed circle is

$$\pi \left(\frac{2 - \sqrt{2}}{2} \right)^2 = \pi \left(\frac{3 - 2\sqrt{2}}{2} \right).$$

Geometric Novelty 28

Problem: The angles of a triangle are in the ratio of 3:4:5. Find the numerical value of the ratio of the sine of the smallest angle to the sine of the next smallest angle.

Solution: Given the ratio of the angles of the triangle, we can represent their relative sizes as $3x$, $4x$, and $5x$. The angle sum is

$$3x + 4x + 5x = 180°,$$

and $x = 15°$. Thus, the angles measure $45°$, $60°$, and $75°$. We seek

$$\frac{\sin 45°}{\sin 60°} = \frac{\sqrt{2}}{\sqrt{3}} = \frac{\sqrt{6}}{3}.$$

These sine functions can easily be obtained from an isosceles right triangle and from an equilateral triangle.

Geometric Novelty 29

Problem: In Figure 4-41, quadrilateral *ABCD* is a trapezoid where $AB = 2CD$, and *E* is the midpoint of *DB*. Also, *GE* intersects *AC* at point *F* and intersects *BA* at point *G* so that $AB = AG$. Find the ratio $\frac{CF}{AF}$.

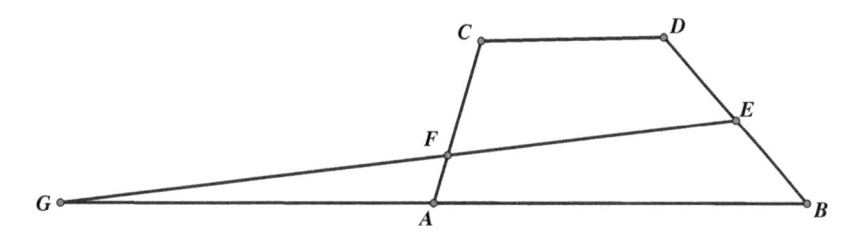

Figure 4-41

Solution: Begin by drawing EH parallel to AB so that point H is the midpoint of AC, as shown in Figure 4-42. Since the length of the median of a trapezoid is the average of the lengths of the bases,

$$EH = \frac{AB + CD}{2}.$$

Also, $AB = 2CD$ so that $EH = \frac{3CD}{2}$. Because $\triangle FAG \sim \triangle FHE$, we get $\frac{FH}{AF} = \frac{EH}{AG}$ and it follows that

$$\frac{FH}{AF} = \frac{EH}{AG} = \frac{\dfrac{3CD}{2}}{2CD} = \frac{3}{4}.$$

Therefore, if $FH = 3a$, then $AF = 4a$, and $AH = 7a = CH$. We then have

$$CF = CH + FH = 10a,$$

and so

$$\frac{CF}{AF} = \frac{10a}{4a} = \frac{5}{2}.$$

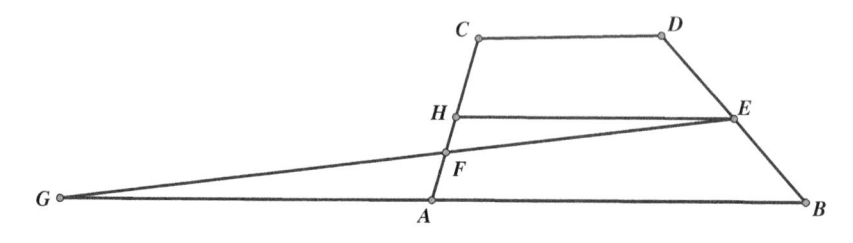

Figure 4-42

Geometric Novelty 30

Problem: In Figure 4-43, isosceles triangle *ABC* has side lengths *AB* = *AC* = 18 and *BC* = 12. Also, altitudes *AE* and *BD* intersect sides *BC* and *AC*, at points *E* and *D*, respectively. Find the length of *DE*.

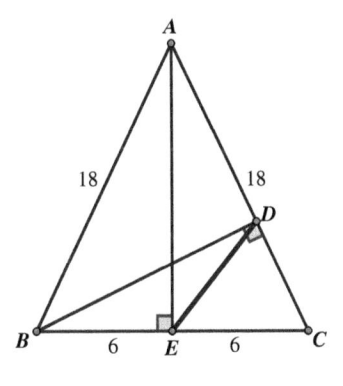

Figure 4-43

Solution: We see immediately that the altitude *AE* bisects side *BC* at point *E*, so that *DE* = *CE* = 6. Consider a circle going through points *B*, *D*, and *C*. In that case, *BC* would be the diameter of the circle, as shown in Figure 4-44, so that *DE* is a radius of the circle and, therefore, has length 6.

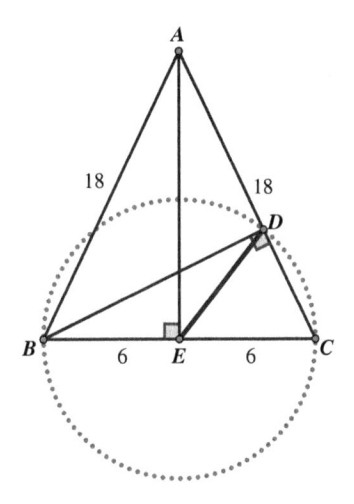

Figure 4-44

Geometric Novelty 31

Problem: From a point *P* outside a circle, secant *PAB* intersects the circle at points *A* and *B*, and secant *PCD* intersects the circle at points *C* and *D*, as shown in Figure 4-45. Lines *BC* and *AD* intersect at point *F.* The angle at point *P* = 70° and ∠*BFD* = 80°. Find the measure of arc *BD*.

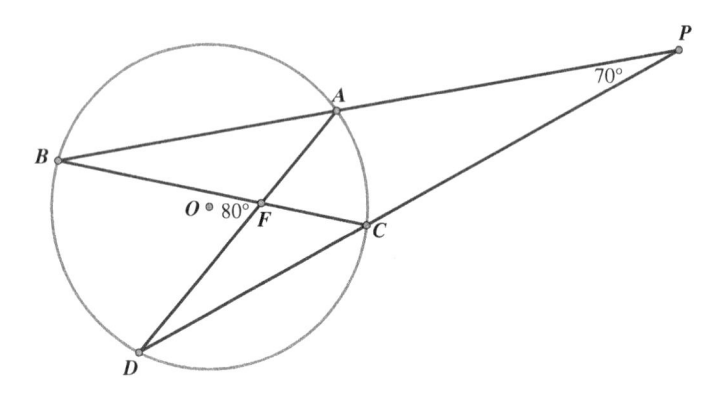

Figure 4-45

Solution: An angle formed by two cords is equal to the measure of the average of the two intersected arcs, which gives us here ∠*BFD* = 80° = $\frac{\widehat{BD}+\widehat{AC}}{2}$. An angle formed by two secants is equal to the measure of half the difference of the intercepted arcs, so we get

$$\angle P = 70° = \frac{\widehat{BD}-\widehat{AC}}{2}.$$

When we add these two equations we get:

$$150° = \frac{\widehat{BD}+\widehat{AC}}{2} + \frac{\widehat{BD}-\widehat{AC}}{2} = \frac{2\widehat{BD}}{2} = \widehat{BD},$$

which is what we sought.

Geometric Novelty 32

Problem: Figure 4-46 shows triangle *ABC* with sides of length 13, 14, and 15. What is the area of the circumscribed circle around triangle *ABC*?

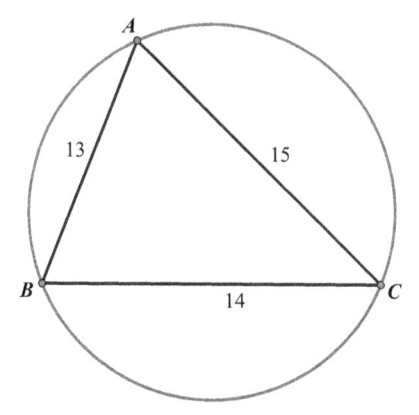

Figure 4-46

Solution: In order to find the area of the circle we need to get the length of the radius. We begin the process by constructing altitude *AD* to side *BC* of triangle *ABC*, as shown in Figure 4-47. The uniqueness of this triangle is such that the length of altitude $AD = 12$, and by the Pythagorean theorem $BD = 5$, and similarly, $DC = 9$. We then consider diameter *AE* which creates right triangle *ACE*, since any angle inscribed in a semicircle is a right angle. We now notice that $\angle B = \angle E$ since both are measured by one-half arc *AC*. Therefore,

$$\triangle ADB \sim \triangle ACE, \text{ and then } \frac{AE}{13} = \frac{15}{12}, \text{ or } AE = \frac{195}{12} = \frac{65}{4}.$$

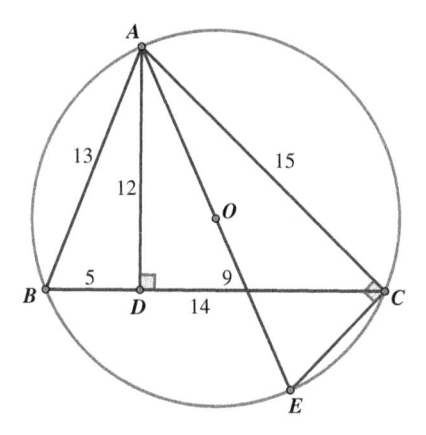

Figure 4-47

To find the area of the circle, radius

$$AO = \frac{1}{2} \cdot \frac{65}{4} = \frac{65}{8},$$

and the area is

$$\pi \left(\frac{65}{8}\right)^2 = \pi \frac{4225}{64} \approx 207.4.$$

Geometric Novelty 33

Problem: In right triangle ABC, side $AC = 3$ and $BC = 4$. The bisector AE of angle BAC intersects side BC at point F and intersects the altitude CD to side AB at point E, as shown in Figure 4-48. Find the ratio $\frac{ED}{CE}$.

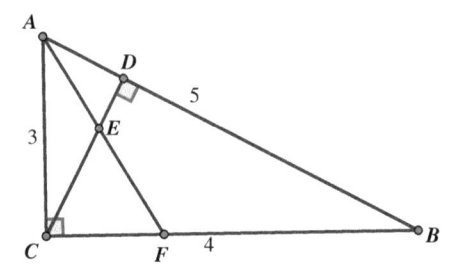

Figure 4-48

Solution: Begin by noting that the bisector of an angle of the triangle partitions the side to which it is drawn proportional to the two adjacent sides of the angle. That provides the following proportion in triangle ABC:

$$\frac{CF}{FB} = \frac{AC}{AB} = \frac{3}{5}.$$

Noticing that $\triangle ABC \sim \triangle ACD$, since the right triangles share a common angle CAB, we then have

$$\frac{AC}{AB} = \frac{AD}{AC} = \frac{3}{5}.$$

Therefore, since AE is an angle bisector in triangle ADC, we get

$$\frac{ED}{CE} = \frac{AD}{AC} = \frac{3}{5},$$

which was sought.

Geometric Novelty 34

Problem: Point A is on side BD of triangle BCD so that $AB = BC$ and $AC = AD$, as shown in Figure 4-49. If $\angle BDC = 20°$, what is the measure of $\angle ABC$?

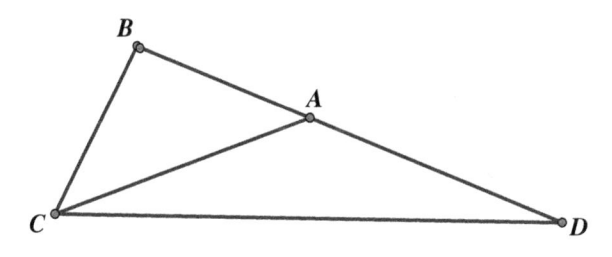

Figure 4-49

Solution: We know that $\angle BDC = 20°$, then for isosceles triangle CAD we have $\angle ACD = 20°$. The exterior angle of triangle CAD is

$$\angle BAC = 20° + 20° = 40°.$$

Then for isosceles triangle ABC,

$$\angle BCA = \angle BAC = 40°.$$

Therefore, in triangle ABC we have

$$\angle ABC = 180° - 80° = 100°.$$

Geometric Novelty 35

Problem: The sides of a given right triangle all have integral lengths and the sum of the lengths of one of the legs and the hypotenuse is equal to 49. Find all possible lengths for the other leg of the right triangle.

Solution: We assume that the legs of the right triangle have lengths a and b, and the hypotenuse has length c. We let $c = b + 49$, or $b = 49 - c$, then apply the Pythagorean theorem

$$a^2 + (49 - c)^2 = c^2.$$

We then get $a = 7\sqrt{2c-49}$ so that $2c - 49$ must equal an odd perfect square less than 49. Our options are 25, 9, and 1, which yield $a = 35, 21$, and 7, respectively.

An alternate solution would be to let the side lengths of the triangle be $m^2 - n^2$, $2mn$, and $m^2 + n^2$, where the first two are the legs of the right triangle and the third is the hypotenuse. Now the sum of one leg and hypotenuse is $m^2 + n^2 + 2mn = (m + n)^2 = 49$, therefore, $m + n = 7$. Thus, the possible values of m and n are (4 and 3), or (5 and 2), or (6 and 1). Which gives us the following values for the length of the other leg, $m^2 - n^2$: 7, 21, and 35.

Geometric Novelty 36

Problem: In Figure 4-50, quadrilateral *ABCD* is a rectangle. What part of the rectangle is the shaded region?

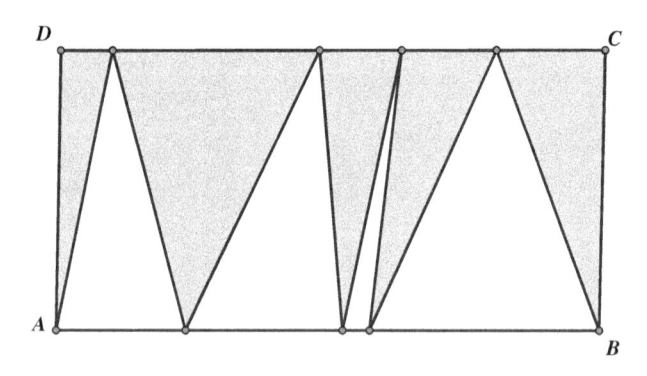

Figure 4-50

Solution: The sum of the areas of the four non-shaded triangles are obtained as follows using the segment lengths denoted in Figure 4-51. That is,

$$\frac{1}{2}hn+\frac{1}{2}hm+\frac{1}{2}hp+\frac{1}{2}hq=\frac{1}{2}h(n+m+p+q)=\frac{1}{2}(AD)(AB),$$

which represents one-half of the area of the rectangle. Therefore, the shaded region is also one-half the area of the rectangle.

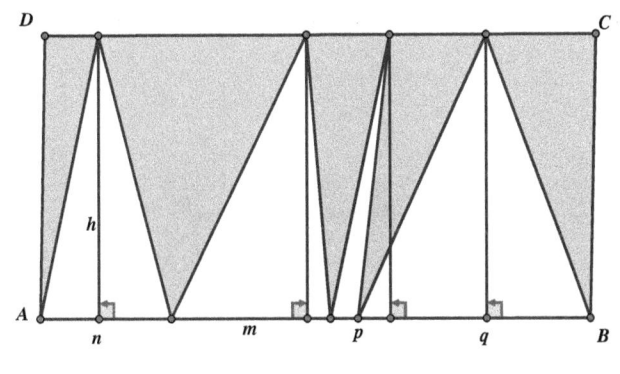

Figure 4-51

Geometric Novelty 37

Problem: Charlie travels 10 miles north, 5 miles west, and finally 2 miles north again to reach his destination. How far is Charlie from his starting point? (See Figure 4-52.)

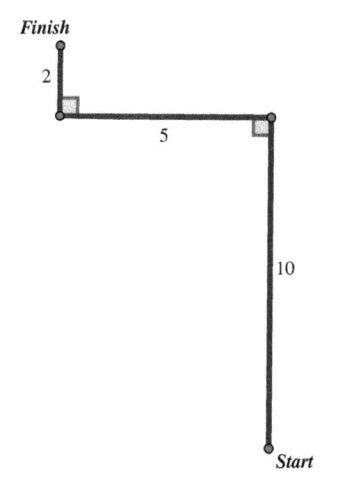

Figure 4-52

Solution: From the path traveled we can create rectangle *ABCD* with dimensions 10 and 5. That is, $AD = 5$ and $DE = CD + CE = 12$, as shown in Figure 4-53. When we draw *AFE* it turns out to be the hypotenuse of triangle *ADE*. By applying the Pythagorean theorem to this triangle, we find that $AE = 13$.

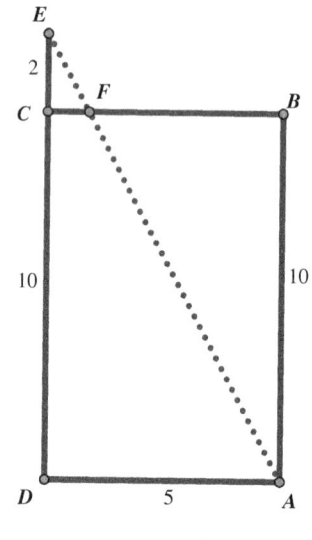

Figure 4-53

Geometric Novelty 38

Problem: In Figure 4-54, rectangle *ABCD* has area 45 and is partitioned into nine congruent rectangles. What is the perimeter of rectangle *ABCD*?

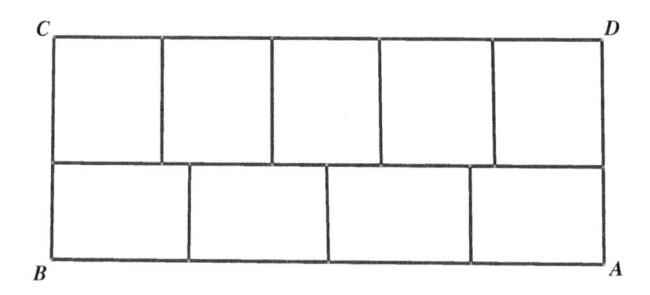

Figure 4-54

Solution: Let w represent the width of each of the small rectangles, and let l represent the length of each of the small rectangles. Since the area of rectangle $ABCD$ is 45, each of the small rectangles has an area of 5. Therefore, $wl = 5$, or $l = \frac{5}{w}$. Furthermore, representing the length of rectangle $ABCD$, we find that $4l = 5w$, or $l = \frac{5w}{4}$. We equate the two representations of l to get

$$\frac{5}{w} = \frac{5w}{4}, \text{ and then } w = 2, \text{ and } l = \frac{5}{w} = \frac{5}{2}.$$

Therefore, the perimeter of rectangle $ABCD$ is

$$(4l + 5w) + 2(l + w) = \left(4 \cdot \frac{5}{2} + 5 \cdot 2\right) + 2\left(\frac{5}{2} + 2\right) = 29.$$

Geometric Novelty 39

Problem: In Figure 4-55, lines AE, AD, and BC are each tangent to circle O at points E, D, and F, respectively. If $AE = 10$, what is the perimeter of triangle ABC?

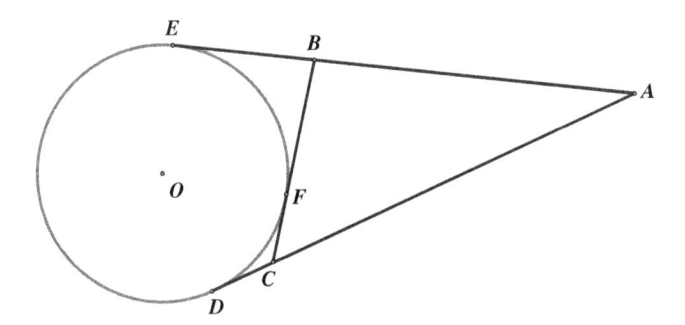

Figure 4-55

Solution: The basic principle to solving this problem is to recognize that from an external point, tangents drawn to the same circle are equal. Therefore, $AE = AD = 10$. Also, $BE = BF$ and $CD = CF$. The perimeter of triangle ABC is

$$(AB + BF) + (AC + CF) = (AB + BE) + (AC + CD) = 10 + 10 = 20.$$

Geometric Novelty 40

Problem: In Figure 4-56, circles with centers B and C intersect at points A and D. The challenge here is to find the perimeter of the shaded region when $BC = 3$.

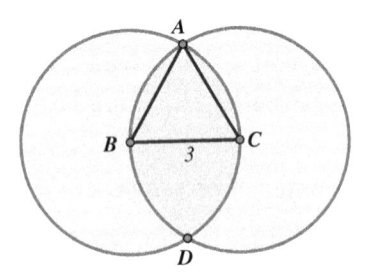

Figure 4-56

Solution: Notice that triangle ABC is an equilateral triangle with sides of length 3. Since $\angle C = 60°$, arc AB is also 60°. Analogously, arc BD is 60° as well. Therefore, the major arc $AD = 240°$, or two-thirds of a circle's circumference. The circumference of each of the two circles is $2\pi \cdot 3 = 6\pi$, and two-thirds of that is 4π. The perimeter of the shaded region is twice two-thirds of the circumference, or 8π.

Geometric Novelty 41

Problem: In Figure 4-57, AO, CO, and DO are radii of circle O. Radius AO extended intersects CD at point B, such that BD is equal to the length of the radius of the circle. If $\angle AOC = 81°$, what is the measure of angle B?

Solution: To simplify matters we will let $\angle B = x$, then for isosceles triangle DOB, base $\angle DOB = x$ and the exterior angle of triangle DOB, namely angle $\angle CDO = 2x$, since the exterior angle of the triangle is equal to the sum of the two remote interior angles. Similarly, for isosceles triangle

COD, we have $\angle C = 2x$. Because angle *AOC* is an exterior angle of triangle *BOC*, we have $\angle AOC = \angle B + \angle C = x + 2x = 3x = 81$, and then $x = 27°$.

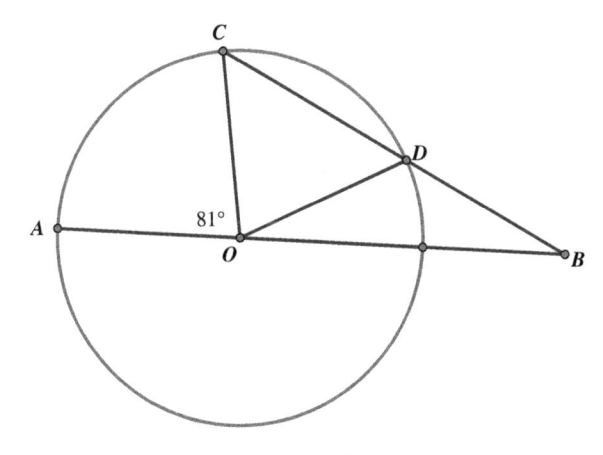

Figure 4-57

Geometric Novelty 42

Problem: Too often we take geometry for granted. Here is a problem that puts some of geometry's basic principles to good use. In Figure 4-58, we have *P* as any point on side *AB* of triangle *ABC*, where *M* is the midpoint of *BC* and *N* is the midpoint of *AC*. What is the ratio of the area of quadrilateral *MCNP* to the area of triangle *ABC*?

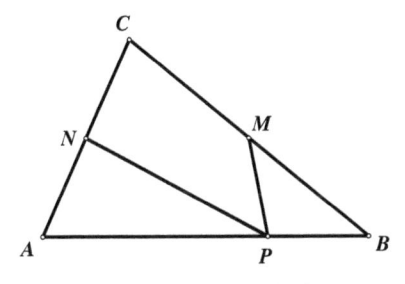

Figure 4-58

Solution: Here is another problem where the elegance emerges in the alternative solution we present after the traditional solution. The expected solution would have us draw *BN*, and then recognize that triangle *ABN* has half the area of triangle *ABC*, since the median of a triangle partitions it into two equal areas, as can be seen in Figure 4-59.

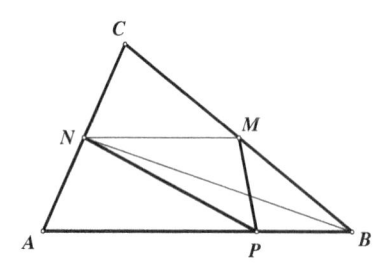

Figure 4-59

Also, since a line joining the midpoints of two sides of a triangle is parallel to the third side, *MN* is parallel to *AB*. Therefore, triangle *BMN* has the same area as triangle *PMN*, because they share the same base *MN* and, since their vertices lie on a line parallel to their common base, have equal altitudes to that base. The area $\triangle BMN$ + area $\triangle CNM$ = area $\triangle BCN = \frac{1}{2}$ area $\triangle ABC$.

By substitution, we have area $\triangle PMN$ + area $\triangle CNM = \frac{1}{2}$ area $\triangle ABC$. That is to say that the area of quadrilateral *MCNP* is one-half of the area of triangle *ABC*.

We could make this problem much easier by choosing a convenient point for *P* on *AB*, since no specific point location for *P* on *AB* was given. Suppose *P* was at one extreme end of *AB* at point *B*. In that case, the quadrilateral *MCNP* reduces to become triangle *BCN*, since the length of *BP* is 0. As we mentioned above, triangle *BCN* results from a median partitioning a triangle; hence, the area of triangle *BCN* is one-half the area of triangle *ABC*.

Geometric Novelty 43

Problem: In Figure 4-60, in triangle *DEF*, we find that *AF* = *DF*, and *DE* = *CE*. If angle *ABC* is 120°, what is the measure of angle *ADC*?

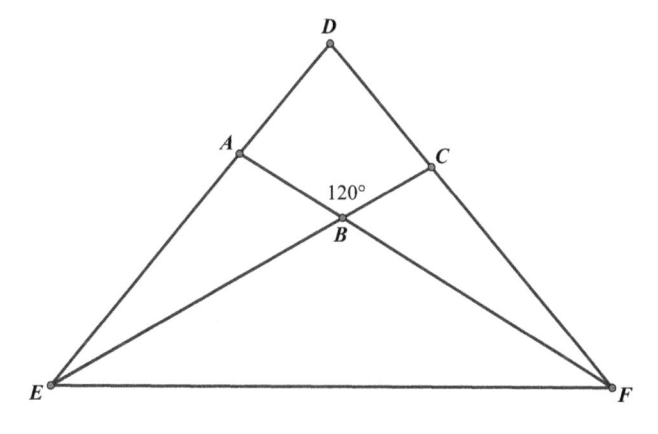

Figure 4-60

Solution: In Figure 4-60, there are two isosceles triangles, $\triangle AFD$, and $\triangle DEC$, each of which has equal base angles. Therefore, $\angle DAF = \angle D$, and $\angle DCE = \angle D$. Since the sum of the angles of quadrilateral $ABCD$ is $360°$, and $\angle ABC = 120°$ there remain $240°$ to be split equally among the remaining three angles of the quadrilateral, therefore, $\angle D = 80°$.

Geometric Novelty 44

Problem: Three circular barrels of equal size are tied together with rope as shown in Figure 4-61. The radius of each of these circular barrels is 1 foot. What is the length of the entire rope circulating these three barrels?

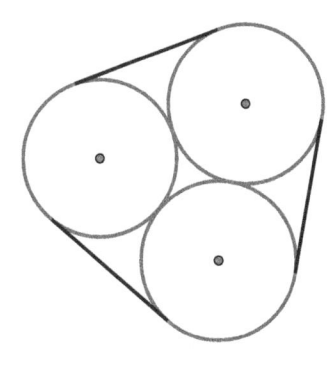

Figure 4-61

Solution: In Figure 4-62, we can clearly see that the three common tangents *DE*, *FG*, and *HJ* have a length equal to twice the radius of the circle, namely, 2 feet in length. Therefore, the combined length of the three tangents is 6 feet. We also notice in Figure 4-62 that the triangle *ABC* is equilateral and, therefore, $\angle BAC = 60°$. Considering the two right angles *FAB* and *EAC*, we find that $\angle FAE = 120°$, or one-third of the circumference of one of the circular barrels. Therefore, the sum of the three arcs *FE*, *DJ*, and *HG* is equal to the circumference of one circle, namely, 2π. Hence, the amount of rope required around these three circular barrels is $6 + 2\pi$.

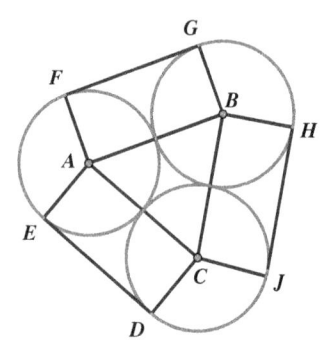

Figure 4-62

Geometric Novelty 45

Problem: In Figure 4-63, the challenge is to find the area of the shaded region formed by two semicircles with diameters 8 and 4, respectively.

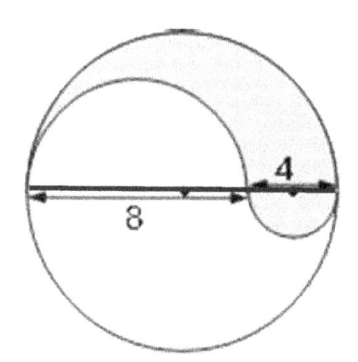

Figure 4-63

Solution: To find the area of the shaded region, subtract the area of the semicircle with diameter 8 from the area of the large semicircle whose diameter is 12. Then we add the semicircle whose diameter is 4. This gives us the shaded region as follows:

$$\frac{1}{2}6^2\pi - \frac{1}{2}4^2\pi + \frac{1}{2}2^2\pi = 18\pi - 8\pi + 2\pi = 12\pi.$$

Geometric Novelty 46

Problem: In Figure 4-64, points A, B, C, D, and E are on the circle. If $\overset{\frown}{AE} = 40°$, the challenge is to find the sum $\angle B + \angle D$.

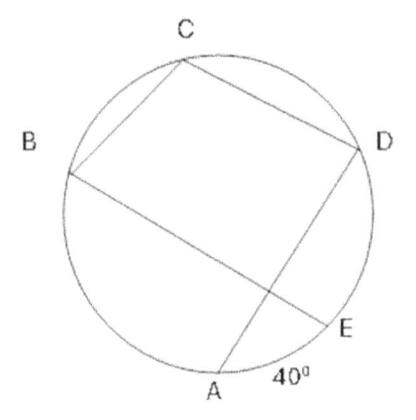

Figure 4-64

Solution: Based on the concept that the inscribed angle in the circle is measured by one-half the intercepted arc, we have the following:

$$\angle D = \frac{1}{2}\overset{\frown}{ABC}, \text{ and } \angle B = \frac{1}{2}\overset{\frown}{EDC}.$$

Therefore, the angle sum

$$\angle D + \angle B = \frac{1}{2}\overset{\frown}{ABCDE} = \frac{1}{2}(360° - 40°) = 160°.$$

Geometric Novelty 47

Problem: How many triangles are in the diagram shown in Figure 4-65?

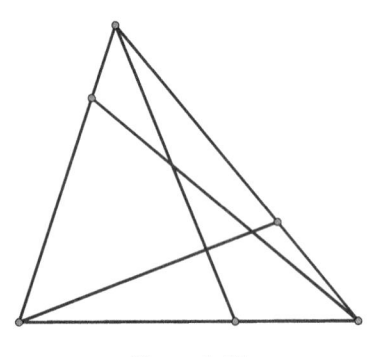

Figure 4-65

Solution: One method to approach this type of problem is to first consider the triangle with only one interior line, where the triangles are $\triangle ABC$, $\triangle ABD$, and $\triangle ADC$, shown in Figure 4-66.

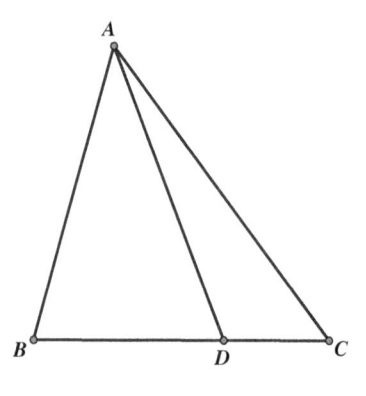

Figure 4-66

Then add another of the interior lines (*BEF*) and count all the triangles that depend on this new line for a side. The various triangles are $\triangle BED$, $\triangle ABE$, $\triangle ABF$, $\triangle AEF$, and $\triangle BFC$, as can be seen in Figure 4-67.

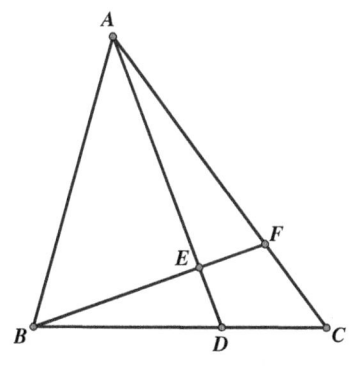

Figure 4-67

Then add the last of the interior lines (*CGHI*), as shown in Figure 4-68, and count all the triangles which depend on it for a side. The various triangles are $\triangle BGC$, $\triangle BIC$, $\triangle HEG$, $\triangle DHC$, $\triangle BIG$, $\triangle AIH$, $\triangle AHC$, $\triangle GFC$, and $\triangle AIC$. In all, there are 17 different triangles.

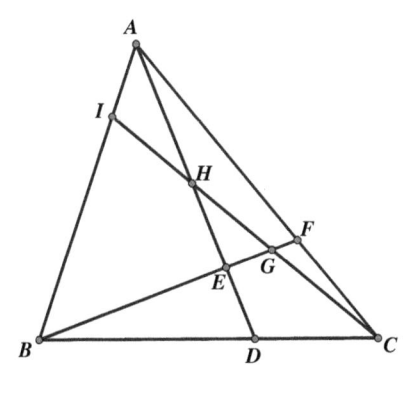

Figure 4-68

Geometric Novelty 48

Problem: Consider a circle whose diameter is increased by π units. By how many units has the circumference been increased as a result of this change in the diameter?

Solution: If D = the diameter of the original circle, then πD is the circumference of the original circle. The new circle's circumference = $\pi(D + \pi) = \pi D + \pi^2$, indicating an increase of π^2 over the original circle's circumference.

Geometric Novelty 49

Problem: In Figure 4-69, the angle measures are marked and the challenge is to measure the angle marked with the "?" symbol.

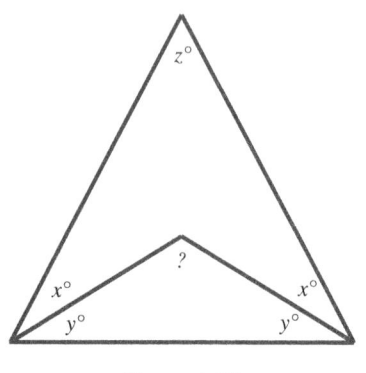

Figure 4-69

Solution: The sum of the angles in the large triangle is $2x + 2y + z$. The sum of the angles in the small triangle is $2y +$?. Since the sum of the angles of all triangles is a constant, $2x + 2y + z = 2y +$?, therefore, $2x + z =$?.

Geometric Novelty 50

Problem: Two circular cans have the same volume, yet their heights are in the ratio of 5:4. What is the ratio of the radii of these two cans?

Solution: Since the cans have equal volumes and the heights of the cans are in the ratio 5:4, the ratio of the areas of their bases is 4:5, because the volume equals the product of the height and the area of the base. If the ratio of their base areas is 4:5, then the ratio of their radii, as well as other corresponding linear parts, is $\sqrt{4} : \sqrt{5} = 2 : \sqrt{5}$.

Geometric Novelty 51

Problem: In Figure 4-70, two circles of equal radii of length 6 contain the center of each other's circle. Find the area where the 2 circles overlap (the darker shaded region).

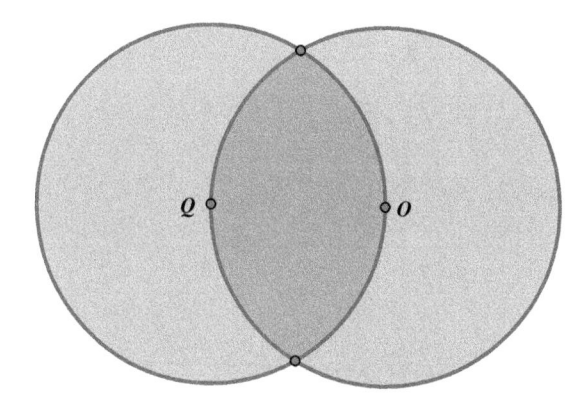

Figure 4-70

Solution: In Figure 4-71, all the straight lines are equal. Therefore, $\triangle OPQ$ and $\triangle ORQ$ are equilateral, and $\angle POQ = \angle ROQ = 60°$. The shaded region is composed of twice the area of sector POQ plus twice the area of segment PQ (i.e., the area of segments PO and RO). The area of sector $POQ = \frac{60°}{360°}\pi(36) = 6\pi$. The area of segment PQ = the area of sector POQ – the area of $\triangle POQ = 6\pi - 9\sqrt{3}$. Hence, the area of the shaded region equals $12\pi + 12\pi - 18\sqrt{3} = 24\pi - 18\sqrt{3}$.

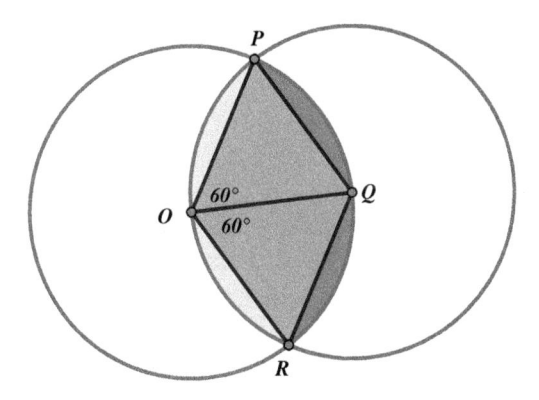

Figure 4-71

Geometric Novelty 52

Problem: The triangle shown in Figure 4-72 has sides of length 13, 14, and 15. The challenge is to find the length of the altitude to the side of length 14.

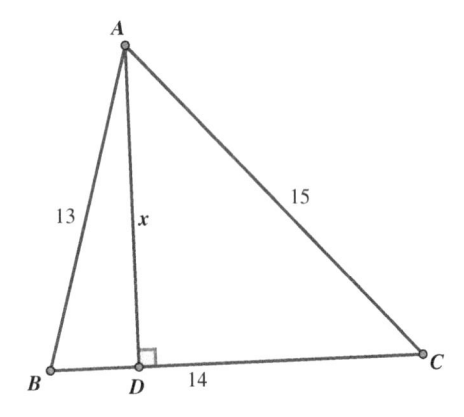

Figure 4-72

Solution: Let x = the length of the altitude to the side of length 14, and let $y = BD$ so that $DC = 14 - y$. By applying the Pythagorean theorem to $\triangle ADB$ we have

$$x^2 + y^2 = 169, \text{ or } x^2 = 169 - y^2. \tag{1}$$

We apply the Pythagorean theorem to $\triangle ADC$: $x^2 + (14 - y)^2 = 225$, then we get

$$x^2 = -y^2 + 28y + 29. \tag{2}$$

By equating lines (1) and (2), we get $169 - y^2 = -y^2 + 28y + 29$, so that $28y = 140$, and $y = 5$. Thus, from equation (1) $x^2 = 169 - y^2 = 169 - 25 = 144$ we have $x = 12$, which is the length of the altitude.

Geometric Novelty 53

Problem: Point E is on side BC of rectangle $ABCD$, as shown in Figure 4-73. Which of these two triangles, $\triangle AEC$ and $\triangle DEC$, is larger?

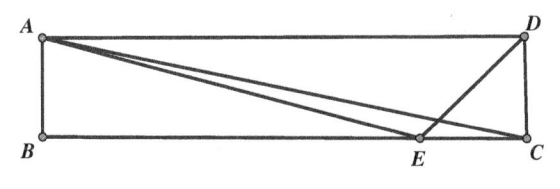

Figure 4-73

Solution: The answer to this question is somewhat camouflaged by the diagram which could be misleading. The answer to the question is rather simple. Since both triangles share the same base EC and have equal altitudes, namely, $AB = DC$, their areas are equal.

Geometric Novelty 54

Problem: Which of the following two areas, shown in Figure 4-74, is greater: the light-shaded region at the four corners of square $ABCD$ or the light-shaded region at the center of the square formed by the four circle arcs?

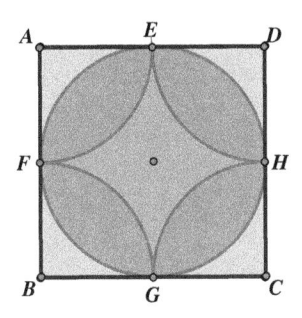

Figure 4-74

Solution: As deceptive as this problem appears, the solution is quite simple. The area at the four corners, which is in light shading, is formed by subtracting the area of the circle from the area of the square. The area at the center, also in light shading, is formed by subtracting the four quarter circles from the area the square. Therefore, the two regions in question are equal in area.

Geometric Novelty 55

Problem: A rectangle is inscribed in a circle and a circle with the same center point is inscribed in the rectangle. The radius of the inscribed circle is 8. Find the area of the rectangle.

Solution: To begin, we notice that the rectangle must be a square, since that is the only shape rectangle that can have a circle inscribed in it, as shown in Figure 4-75. Therefore, the side of the square is 16, and thus, the area of the square is $16^2 = 256$.

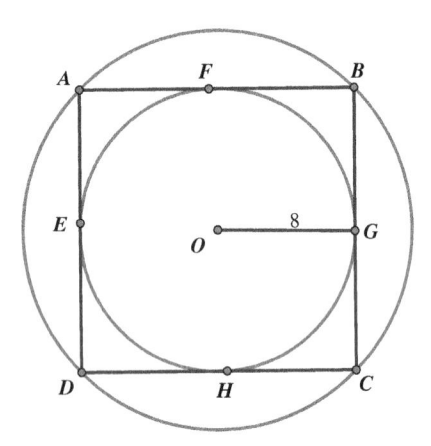

Figure 4-75

Geometric Novelty 56

Problem: In Figure 4-76, the length of the sides of triangle *ABC* are as follows: $AB = 425$, $BC = 450$, and $CA = 510$. The three line segments *HJ*,

GF, and *DE* are all the same length *x*, and intersect at point *P*. Furthermore, the lines *HJ*, *GF*, and *DE* are parallel to the sides of triangle *ABC*, respectively. Find the value of *x*.

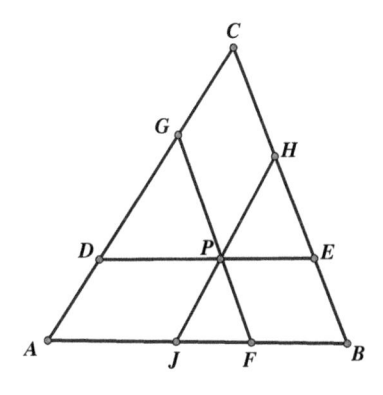

Figure 4-76

Solution: Because of the parallel lines *HJ*||*AC*, *GF*||*BC*, and *DE*||*AB*, we have three parallelograms: *ADPJ*, *BEPF*, and *CGPH*. This enables us to get the following: $EH = BC - (BE + HC) = BC - (FP + GP) = 450 - x$. Similarly, we get $DC = 510 - x$.

We also have similar triangles: $\triangle GPD \sim \triangle ABC$, so that $\frac{PD}{GD} = \frac{AB}{AC}$. Therefore,

$$PD = GD\frac{AB}{AC} = \frac{425}{510}(510 - x) = 425 - \frac{5}{6}x.$$

Analogously we get $\triangle PEH \sim \triangle ABC$, so that $\frac{PE}{HE} = \frac{AB}{BC}$. Therefore,

$$PE = \frac{AB}{BC} \cdot HE = \frac{425}{450}(450 - x) = 425 - \frac{17}{18}x.$$

Since

$$x = PD + PE = \left(425 - \frac{5}{6}x\right) + \left(425 - \frac{17}{18}x\right) = 850 - \frac{16}{9}x,$$

whereupon $x = 306$.

Geometric Novelty 57

Problem: Consider the globe of the earth with a rope wrapped tightly around the equator. The rope will be about 24,900 miles long. We now lengthen the rope by exactly 1 yard. We position this (now loose) rope around the equator so that it is uniformly spaced off the globe, as shown in Figure 4-77. Will a mouse fit under the rope?

Figure 4-77

Solution: The traditional way to determine the distance between the circumferences is to find the difference between the radii. Let R be the length of the radius of the circle formed by the rope (circumference $C + 1$) and r the length of the radius of the circle formed by the earth (circumference C), as shown in Figure 4-78.

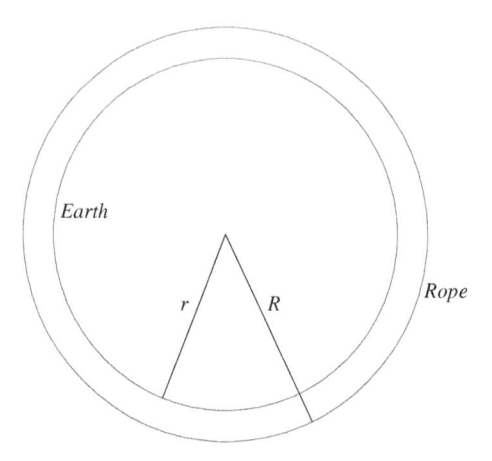

Figure 4-78

The circumference formula gives us: $C = 2\pi r$, or $r = \frac{C}{2\pi}$. Also $C + 1 = 2\pi R$, or $R = \frac{C+1}{2\pi}$.

We need to find the difference of the radii, which is:

$$R - r = \frac{C+1}{2\pi} - \frac{C}{2\pi} = \frac{1}{2\pi} \approx .159 \text{ yards} \approx 5.7 \text{ inches.}$$

Thus, there is a space of over $5\frac{1}{2}$ inches for a mouse to crawl under. One should really appreciate this astonishing result: lengthening the 24,900-mile rope by 1 yard lifted it off the equator about $5\frac{1}{2}$ inches!

Now for an even more elegant solution, consider extreme cases. You should realize that the solution to the problem is independent of the circumference of the earth, since the end result did not include the circumference in the calculation. It only required calculating $\frac{1}{2\pi}$.

Suppose the inner circle (above) is very small, so small that it has a zero-length radius (that means it is actually just a point). This is the extreme case. We seek to find the difference between the radii, $R - r = R - 0 = R$. So, all we need to find is the length of the radius of the larger circle and our problem will be solved. With the circumference of the smaller circle now 0, we apply the formula for the circumference of the larger circle: $C + 1 = 0 + 1 = 2\pi R$, then $R = \frac{1}{2\pi}$. Using the extreme case solved the problem immediately.

Geometric Novelty 58

Problem: Find the measure of the angle ABC formed by the two diagonals of the given cube shown in Figure 4-79.

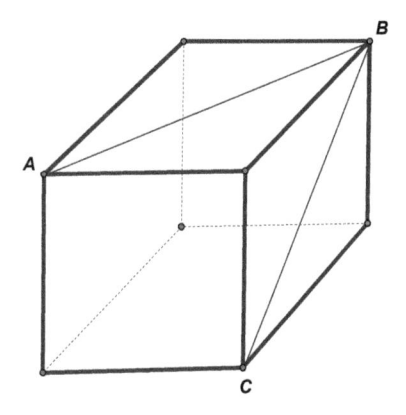

Figure 4-79

Solution: A first reaction is to somehow work with the trihedral angle at *B*. This will lead to a great deal of confusion. If one adopts a different point of view, then one could focus on the plane determined by points *A*, *B*, and *C*. By drawing the segment *AC*, (see Figure 4-80) which has the same length as the other two given diagonals *AB* and *BC*, we get and equilateral triangle *ABC*, where each angle is 60°. Thus, $\angle ABC = 60°$.

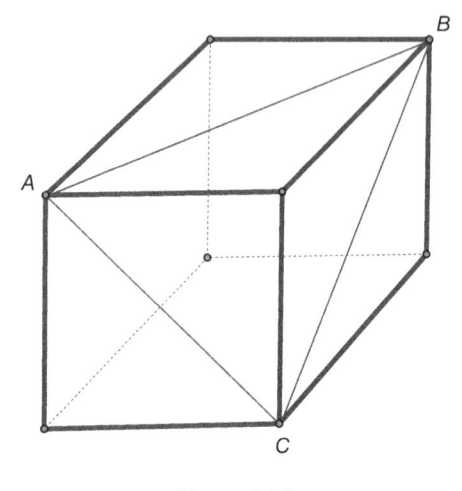

Figure 4-80

Geometric Novelty 59

Problem: In Figure 4-81 we have nine wheels touching each other with diameters successively increasing by 1 cm. Beginning with 1 cm as the smallest circle, and 9 cm for the largest circle, how many degrees does the largest circle turn when the smallest circle turns by 90°?

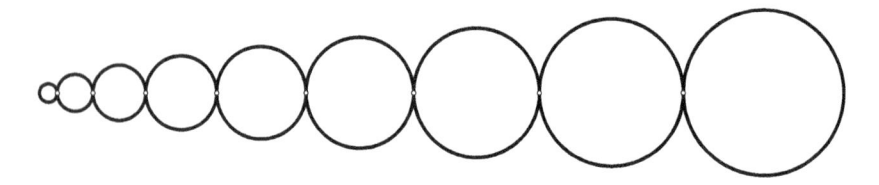

Figure 4-81

Solution: A point on the circumference of a circle with diameter d that rotates α degrees will turn $\frac{\alpha}{360} \cdot \pi d$, where the circumference is πd. To determine the motion that we require here of α degrees, we equate the motion of the smallest circle with that of the largest circle:

$$\frac{90}{360} \cdot \pi \cdot 1 = \frac{\alpha}{360} \pi \cdot 9,$$

which gives us $\alpha = 10°$.

Geometric Novelty 60

Problem: In Figure 4-82, point M is the midpoint of side AB of $\triangle ABC$. Point P is any point on AM. The line through point M, a line parallel to PC, meets BC at D. What part of the area of $\triangle ABC$ is the area of $\triangle BDP$?

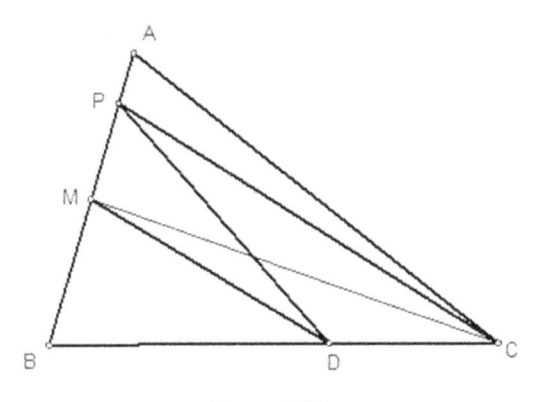

Figure 4-82

Solution: The area of $\triangle BMC$ is one-half the area of $\triangle ABC$ (since the median partitions a triangle into two equal areas).

area$\triangle BMC$ = area$\triangle BMD$ + area$\triangle CMD$ = area$\triangle BMD$ + area$\triangle MPD,$

which equals area

$$\triangle BPD = \frac{1}{2} \text{ area} \triangle ABC.$$

This rests on the property that when the vertices of two triangles lie on a line parallel to a common base, their areas are equal.

The problem can be simplified by using the strategy of considering extreme cases. Let us select point P at an extreme position, either at point M or point A. Suppose P is selected at point A. Notice that as P moves along AB towards point A, line MD, which must stay parallel to PC, moves towards a position whereby D approaches the midpoint of BC. That final position for D then has AD as a median of $\triangle ABC$. Thus, the area of $\triangle PBD$ is one-half the area $\triangle ABC$.

For this extreme case, we must watch all movements when we move a point to an extreme position. This is a very useful example and contrasts nicely to the problem's previous solution.

Geometric Novelty 61

Problem: Here we are given five circles tangent to one another and placed as shown in Figure 4-83. The problem here is to find a line through the center point of the left-most circle that will equally divide the areas of these five circles.

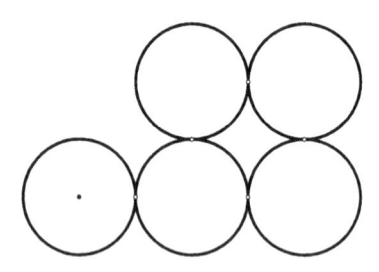

Figure 4-83

Solution: Without an auxiliary line this problem would be difficult to solve. In Figure 4-84, we add three additional circles to the diagram so as to have an even number of circle columns, which will aid our solution. We now draw this auxiliary line through the center of the lower left circle and containing the center of the upper right circle. We have thereby partitioned these circles into two equal area parts. In particular, the five original circles are now partitioned equally. For the curious reader, we offer the fact that the angle α that the auxiliary line made with the horizontal is

$$\alpha = \arctan\frac{1}{3} \approx 18.43°.$$

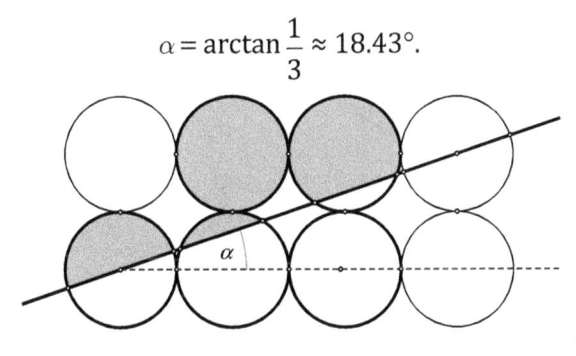

Figure 4-84

Geometric Novelty 62

Problem: What is the radius of the inscribed circle of a triangle whose perimeter is numerically equal to its area?

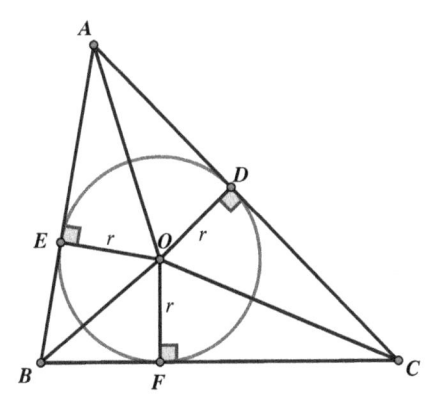

Figure 4-85

Solution: In Figure 4-85, circle O, with radius r, is inscribed in triangle ABC. The area of triangle ABC is

$$\text{area}\triangle AOB + \text{area}\triangle BOC + \text{area}\triangle AOC$$

$$= \frac{1}{2}r(AB) + \frac{1}{2}r(BC) + \frac{1}{2}r(AC) = \frac{1}{2}r(AB + BC + AC).$$

Therefore, the area of triangle ABC is $\frac{1}{2}r(\text{perimeter})$. Thus, when the circle's radius is 2, the triangle's area will be numerically equal to its perimeter.

Geometric Novelty 63

Problem: In Figure 4-86, secants PFC and PEB are drawn for circle O. The point A is on arc BC so that $\overset{\frown}{AC} = 40°$, and $\overset{\frown}{AB} = 70°$. What is the sum of the angle measures: $\angle P + \angle FAE$?

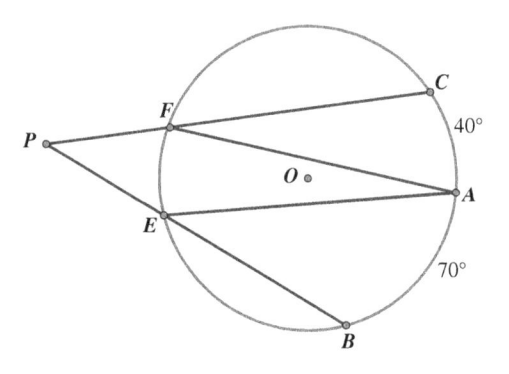

Figure 4-86

Solution: The measurements of the inscribed angles are $\angle AFC = \frac{1}{2}40° = 20°$, and $\angle AEB = \frac{1}{2}70° = 35°$. Their supplements are $\angle AFP = 180° - 20° = 160°$, and $\angle AEP = 180° - 35° = 145°$, respectively. Since the sum of the angles of a quadrilateral is $360°$, the sum of angles P and FAE must equal $360° - 160° - 145° = 55°$.

Geometric Novelty 64

Problem: Points D and E are trisection points on sides AC and AB of equilateral triangle ABC, as shown in Figure 4-87. Also, CE and BD intersect at point P. Find the measure of angle APC.

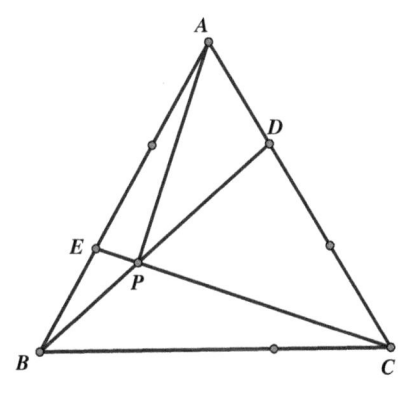

Figure 4-87

Solution: Since $AC = BC$ and AE is two-thirds of the side of the equilateral triangle ABC, we have $AE = DC$. Furthermore, we have $\angle EAC = \angle DCB = 60°$, so that we then have $\triangle EAC \cong \triangle DCB$, and $\angle AEC = \angle CDB$. Angle PDA is supplementary to angle PDC, and therefore, angle PEA is supplementary to angle PDA, which makes quadrilateral $AEPD$ a cyclic quadrilateral. Draw a line ED as shown in Figure 4-88, where we then have $AE = 2AD$ in triangle ADE. Since $\angle EAD = 60°$, we have triangle ADE as a $30° - 60° - 90°$ triangle, so that we have $\angle ADE = 90°$. This establishes that AE is the diameter of the circumscribed circle about quadrilateral $AEPD$. Thus, we have angle $EPA = 90°$ which allows us to conclude that angle APC equals $90°$.

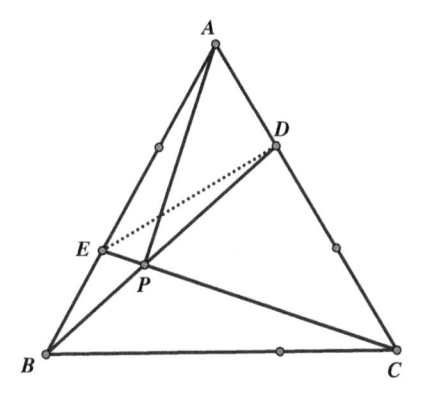

Figure 4-88

Geometric Novelty 65

Problem: Equilateral triangles *DAB* and *FAC* are drawn on sides *AB* and *AC*, respectively, of triangle *ABC*. Point *G* is a centroid of triangle *AFC* and point *M* is the midpoint of side *BC*, as shown in Figure 4-89. What is the measure of angle *DMG*?

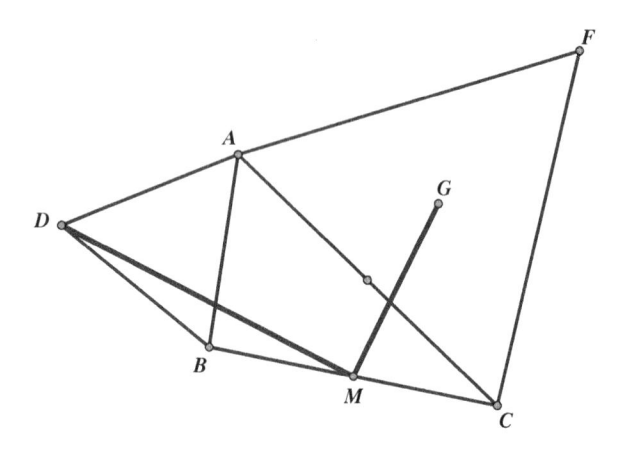

Figure 4-89

Solution: As shown in Figure 4-90, draw the perpendicular from centroid *G* to intersect side *AC* at midpoint *N*. This establishes $\angle AGN = 60°$ so that triangle *AGN* is a $30° - 60° - 90°$ triangle, and then $2GN = AG$. To show that $\angle DAG = \angle MNG$, we note that *MN* is parallel to *AB*, since *MN* is a line joining the midpoints of two sides of triangle *ABC*, so that we have $\angle MNC = \angle BAC$. Therefore, $\angle MNG = 90° + \angle MNC$ and $\angle DAG = \angle BAC + \angle DAB + \angle NAG = \angle BAC + 60° + 30° = \angle MNC + 90° = \angle MNG$. Since $MN = \frac{1}{2}AB = \frac{1}{2}AD$, we have $\triangle MNG \sim \triangle DAG$, and $\frac{GM}{GD} = \frac{1}{2}$. We can now show that $\angle DGM = 60°$ since $\angle AGD + \angle DGN = 60°$ so that $\angle DGM = \angle NGM + \angle DGN = 60°$. This establishes triangle *DGM* is a $30° - 60° - 90°$ triangle and, therefore, $\angle DMG = 90°$.

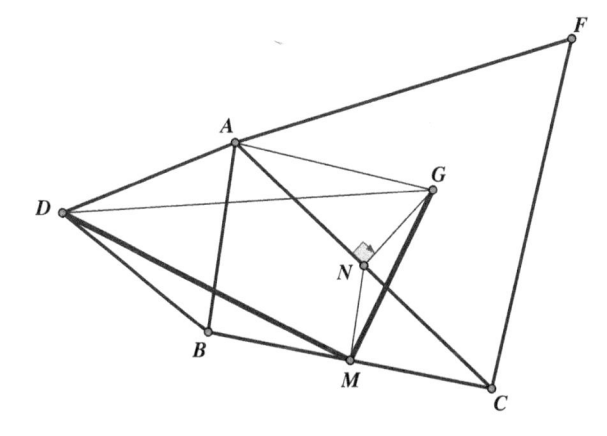

Figure 4-90

Geometric Novelty 66

Problem: In Figure 4-91, $AC = 6$, $BC = 4$, $BD = 7.5$, and $AB = 5$. Also, $\angle ACB = \angle ABD$. Find the length of AD.

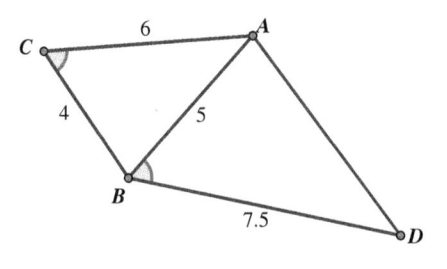

Figure 4-91

Solution: Because two sides of triangle ABC are proportional to two sides of triangle ABD and the included angles $\angle C$ and $\angle ABD$ are equal, the two triangles are similar, and therefore, we have the following proportion:

$$\frac{AC}{BC}=\frac{BD}{AB}, \text{ since } \frac{6}{4}=\frac{7.5}{5}.$$

Thus, we also have the following proportion:

$$\frac{AD}{7.5}=\frac{5}{6}, \text{ so that } AD=6.25.$$

Geometric Novelty 67

Problem: In a triangle, the three medians are 9, 12, and 15 units long. Find the unit length of the side to which the longest median is drawn.

Solution: Triangle ABC has medians BD, CE, and AF, of lengths 9, 12, and 15, respectively, as shown in Figure 4-92. We seek the length of BC.

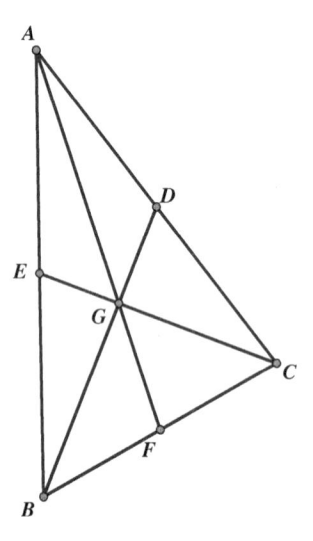

Figure 4-92

Since the medians of the triangle trisect each other, we have $BG = 6$, $CG = 8$, and $GF = 5$. We then extend GF its own length to point K, so that $GK = 10$, as can be seen in Figure 4-93. Because the diagonals in quadrilateral *BGCK* bisect each other, it is a parallelogram and, therefore, *BK* = 8. The sides of triangle *GBK* are 6, 8, and 10, and thus, by the Pythagorean theorem it is a right triangle. Therefore, quadrilateral *BGCK* is a rectangle and its diagonals are equal, so that *GK* = *BC* = 10, which is what we originally sought.

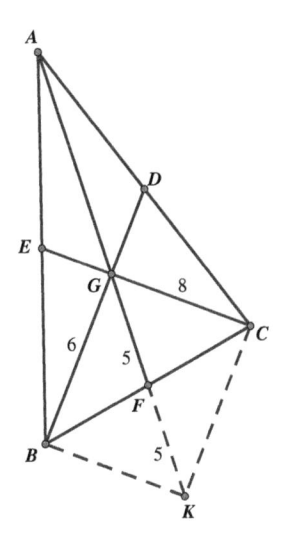

Figure 4-93

Geometric Novelty 68

Problem: Isosceles triangle *ABC* in Figure 4-94 has *AB* = *AC*, altitude *AD* with length 6, and a perimeter of 18. Find the area of triangle *ABC*.

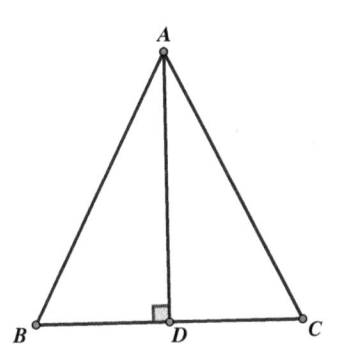

Figure 4-94

Solution: We can express the perimeter of triangle ABC as $2AB + 2BD = 18$, so that $AB + BD = 9$. Applying the Pythagorean theorem to right triangle ABD, we have $AB^2 - BD^2 = 36 = (AB + BD)(AB - BD) = (9)(AB - BD)$. Therefore, $AB - BD = 4$, and subtracting this from the previous $AB + BD = 9$, we get $2BD = 5 = BC$. This enables us to find the area of triangle ABC as

$$\frac{1}{2}(AD)(BC) = \frac{1}{2} \cdot 6 \cdot 5 = 15.$$

Geometric Novelty 69

Problem: In Figure 4-95, point A is on circle O and point B is on radius OBE such that $OB = 6$ and $BE = 4$. Also, point C is on radius OF and quadrilateral $ABOC$ is a rectangle. Find the length of BC.

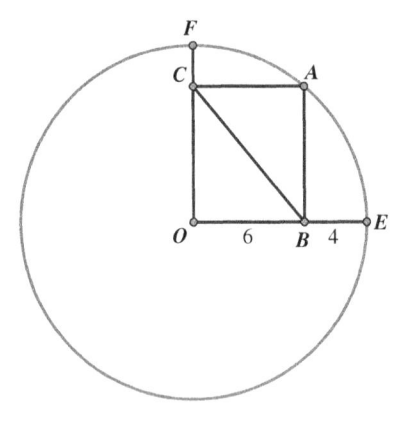

Figure 4-95

Solution: This is rather simple, since $ABOC$ is a rectangle where the diagonals are equal and where, in this case, $AO = EO = 10$. Therefore, diagonal $BC = 10$.

Geometric Novelty 70

Problem: In Figure 4-96, we have isosceles triangle ACD, where $AC = AD$ and point B is on CD so that $AB = AE$. Also, $\angle BAC = 40°$. Find the measure of angle DBE.

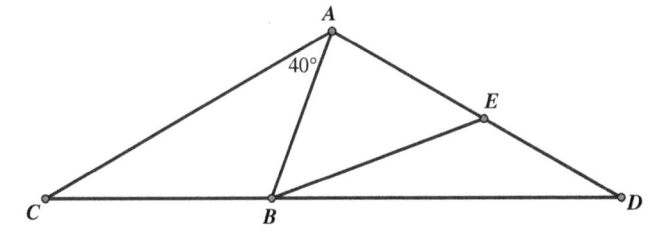

Figure 4-96

Solution: To simplify matters we will mark the angles as shown in Figure 4-97, where $\angle ACD = \angle ADC = p$, and $\angle ABE = \angle BEA = q$, and we seek $\angle DBE = x$. In triangle BDE we have $q = x + p$, and in triangle ABC we have $x + q = p + 40°$. When we replace q to get $x + (x + p) = p + 40°$, we then get $x = 20° = \angle DBE$.

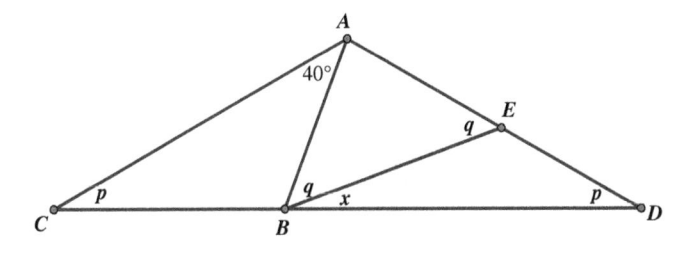

Figure 4-97

Geometric Novelty 71

Problem: In Figure 4-98, point D is on side AB of triangle ABC, and DE is parallel to side AC, and also bisects angle BDC. Furthermore, $\angle BDC = 60°$ and $DC = 6$. Find the length of AC.

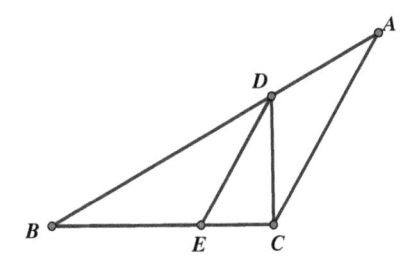

Figure 4-98

Solution: Since $\angle BDC = 60°$ and DE bisects angle BDC, we have $30° = \angle CDE = \angle EDB = \angle A = \angle ACD$ resulting from the parallel lines. As shown in Figure 4-99, we draw $DF \perp AC$, so that $\angle CDF = 60°$. Because triangle DFC is a $30° - 60° - 90°$ triangle, we have $CF = 3\sqrt{3}$, and thus, $AC = 2CF = 6\sqrt{3}$.

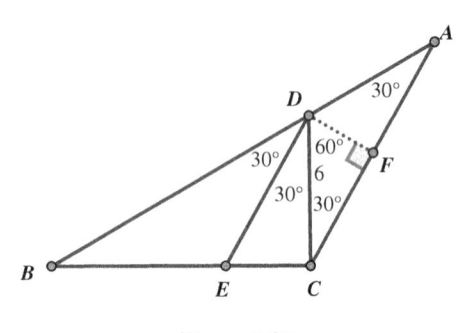

Figure 4-99

Geometric Novelty 72

Problem: In Figure 4-100, triangle ABC has angle bisector BG, and exterior angle ACF of triangle ABC is bisected by CG. Also, DG is parallel to BF intersecting AC at point E. If $BD = 8$ and $CE = 6$, find the length of ED.

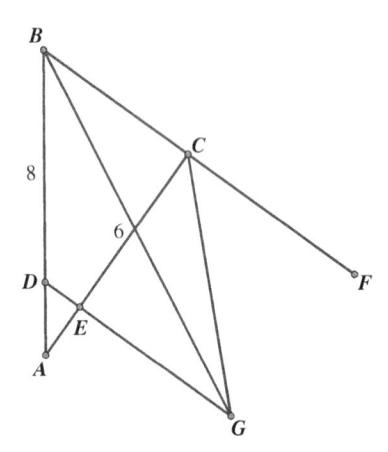

Figure 4-100

Solution: Since $DG \| BC$, we have $\angle EGC = \angle FCG = \angle GCE$, as well as $\angle DGB = \angle CBG = \angle GBD$. This establishes triangle BDG as isosceles so that $DG = BD = 8$. Also, since triangle CEG is isosceles, we have $EG = EC = 6$.

Therefore, $ED = DG - EG = 8 - 6 = 2$.

Geometric Novelty 73

Problem: In Figure 4-101, the area of triangle ABC is 360 and the points D, E, and M are the midpoints of the sides of the triangle. Lines DM and BE intersect at point F, and lines EM and CD intersect at point H. Medians BE and CD intersect at point G. Find the area of triangle DGF.

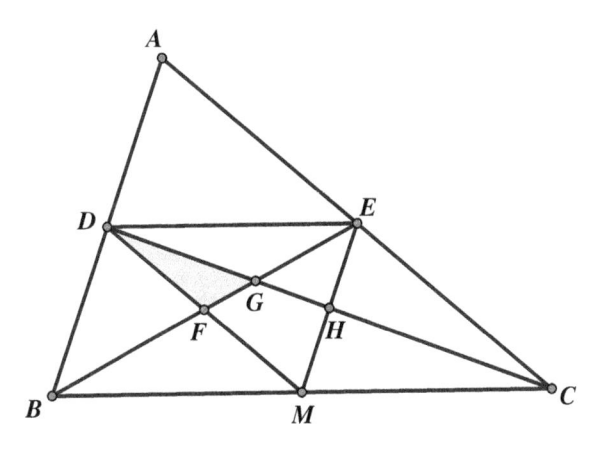

Figure 4-101

Solution: Since triangles DBM, CEM, ADE, and DEM are equal in area, we have

$$\text{area}\triangle DEM = \frac{1}{4}\text{area}\triangle ABC = 90.$$

In Figure 4-102, the three medians of triangle ABC are also the three medians of triangle DEM. Since the medians partition a triangle into six equal areas,

$$\text{area}\triangle DGF = \frac{1}{6}\text{area}\triangle DEM = \frac{1}{6}\cdot 90 = 15.$$

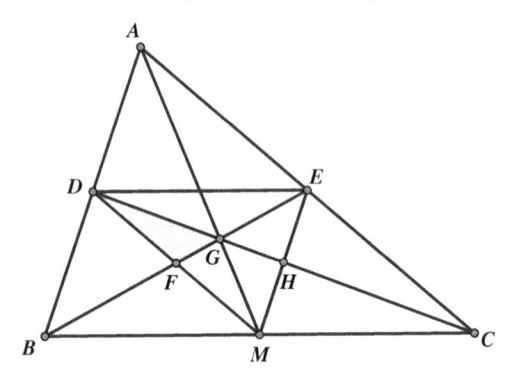

Figure 4-102

Geometric Novelty 74

Problem: In Figure 4-103, within triangle ABC are the following ratios:

$$\frac{AD}{DC} = \frac{5}{1} \quad \text{and} \quad \frac{BE}{ED} = \frac{5}{1}.$$

If the area of triangle ABC is 360, what is the area of triangle AED?

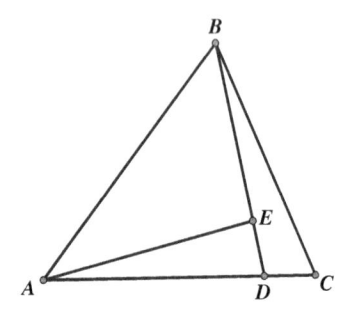

Figure 4-103

Solution: Since triangle *ABD* and triangle *ABC* share a common altitude *BF*, as shown in Figure 4-104, the

$$\text{area}\triangle ABD = \frac{5}{6}\text{area}\triangle ABC = \frac{5}{6}\cdot 360 = 300.$$

Furthermore, since triangle *AED* and triangle *ABD* share a common altitude *AG*, we have the following area relationship:

$$\text{area}\triangle AED = \frac{1}{6}\text{area}\triangle ABD = \frac{1}{6}\cdot 300 = 50.$$

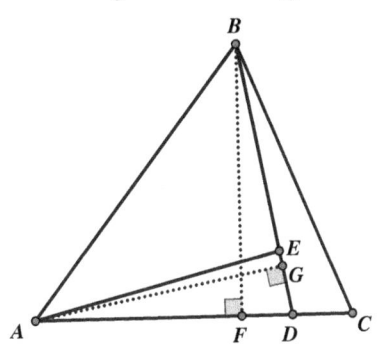

Figure 4-104

Geometric Novelty 75

Problem: In Figure 4-105, equilateral triangles *ADC* and *AEB* are drawn on the sides of triangle *ABC*. If the length of *EC* = 10, what is the length of *BD*?

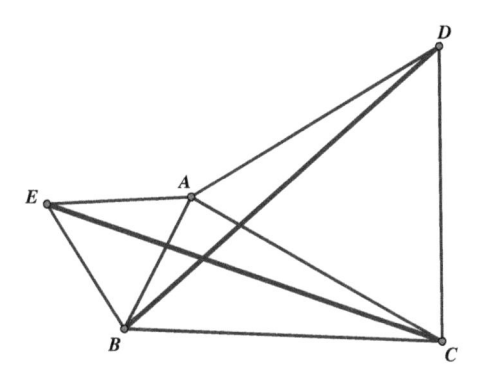

Figure 4-105

Solution: In Figure 4-106, $AE = AB$ and $AD = AC$. Furthermore, $\angle EAB = \angle DAC = 60°$, so that $\angle EAC = \angle BAC + 60° = \angle BAD$. We have $\triangle AEC \cong \triangle ABD$, and therefore, $BD = EC = 10$.

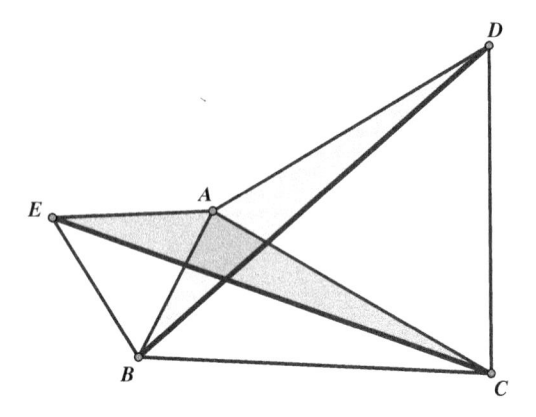

Figure 4-106

Chapter 5

Neglected Topics in Algebra

Although the basic skills of algebra are taught in secondary schools, there are still many aspects of algebra that are either neglected or omitted because of lack of time. These topics provide further insights into problem-solving and enable a better understanding of unusual number relationships. Here we will explore various algebra related topics that could easily have been included in secondary schools if time and interest were available. For example, the first topic that we will explore is an unusual situation, where one equation with two variables offers a new, alternate solution.

Diophantine Equations: Linear Equations with Two Variables

In secondary school, we are taught to expect a second equation when given one equation (usually linear) with two variables. This is quite normal, for when a unique solution is sought two equations are usually needed. This should not imply that one equation with two variables is unsolvable. To the contrary: one equation with two variables can be graphed on the xy-plane. This should confirm that there are an infinite number of (x, y) values which satisfy the equation—the points on the line. Yet, suppose the question requires integer solutions. This leads to some interesting strategies, neglected in the school algebra curriculum, for approaching linear equations.

Suppose you are given 5 dollars and asked to go to the post office to buy 6-cent and 8-cent stamps. How many combinations of 6-cent and 8-cent stamps could you select from to make your purchase? You may realize that there are two variables which must be determined, say x and y. Let x represent the number of 8-cent stamps and y represent the number of 6-cent stamps, and the equation: $8x + 6y = 500$ should follow, which can then be written as $4x + 3y = 250$. At this juncture you may realize that although this equation has infinite number of solutions, it may or may not have an infinite number of *integral* solutions; moreover, it may or may not have an infinite number of *positive* integral solutions, as called for by the original problem of buying stamps. The first issue to consider is whether integral solutions, in fact, exist.

For this, a useful theorem may be employed. It states that "if the greatest common factor of a and b is also a factor of k, where a, b, and k are integers, then there exists an infinite number of integral solutions for x and y for $ax + by = k$." Equations of this type, whose solutions must be integers, are known as Diophantine equations in honor of the Greek mathematician Diophantus of Alexandria (c. 200–c. 284), who wrote about them.

Since the greatest common factor of 3 and 4 is 1, which is a factor of 250, there must exist an infinite number of integral solutions to the equation $4x + 3y = 250$. The question now is how many (if any) *positive* integral solutions exist? One possible method of solution is often referred to as Euler's method, as it was developed by the Swiss mathematician Leonhard Euler (1707–1783). To begin, we should solve for the variable with the coefficient of least absolute value, in this case, y. Thus, $y = \frac{250-4x}{3}$. We rewrite this equation to separate the integral parts: $y = 83 + \frac{1}{3} - x - \frac{x}{3} = 83 - x + \frac{1-x}{3}$. We then introduce another variable, say t, and let $t = \frac{1-x}{3}$. Solving for x yields $x = 1 - 3t$. Since there is no fractional coefficient in this, the process does not have to be repeated as it otherwise would have to be (i.e., each time introducing new variables, as with t above). Substituting for x in the above equation yields $y = \frac{250-4(1-3t)}{3} = 82 + 4t$. For various integral values of t, corresponding values for x and y will be generated. A table of values (Figure 5-1) might prove useful.

t	...	-2	-1	0	1	2	...
x	...	7	4	1	-2	-5	...
y	...	74	78	82	86	90	...

Figure 5-1

Perhaps by extending the table, we would notice what values of t generate positive integral values for x and y. However, this procedure for determining the number of positive integral values of x and y is not very elegant. We turn instead to the following inequalities, which can be solved simultaneously: $1 - 3t > 0$, and $82 + 4t > 0$. Thus, $t < \frac{1}{3}$, and $t > -20\frac{1}{2}$, or $-20\frac{1}{2} < t < \frac{1}{3}$. This indicates that there are 21 possible combinations of 6-cent and 8-cent stamps which can be purchased for 5 dollars.

It might be helpful to observe the solution of a more difficult Diophantine equation, such as the following example. Solve the Diophantine equation $5x - 8y = 39$. We begin the solution by solving for x, since its coefficient has the lower absolute value of the two coefficients $x = \frac{8y+39}{5} = y + 7 + \frac{3y+4}{5}$. We then let $t = \frac{3y+4}{5}$, and solve for y to get $y = \frac{5t-4}{3} = t - 1 + \frac{2t-1}{3}$. Next, we let $u = \frac{2t-1}{3}$, then solve for t since we still have a fraction to contend with. Let $v = \frac{u+1}{2,}$ and then solve for u to get $u = 2v - 1$.

We may now reverse the process since the coefficient of v is an integer. Substituting in the reverse order: $t = \frac{3u+1}{2}$. Therefore, $t = \frac{3(2v-1)+1}{2} = 3v - 1$, and also $y = \frac{5t-4}{3}$. Thus, $y = \frac{5(3v-1)-4}{3} = 5v - 3 = y$. Similarly, $x = \frac{8y+39}{5}$. Therefore, $x = \frac{8(5v-3)+39}{5} = 8v + 3 = x$, so that we now have both x and y in terms of v. We can then see in Figure 5-2 that for integral values of $v > 0$, positive values of x and y will be generated, which was our original goal.

v	...	-2	-1	0	1	2	...
x	...	-13	-5	3	11	19	...
y	...	-13	-8	-3	2	7	...

Figure 5-2

Summing Power Series

Many formulas are often given without proper explanation or justification. Here we will provide a technique for deriving common formulas for summing a series of powers. We begin by finding the sum of the natural number arithmetic progression: $1 + 2 + 3 + \cdots + n$.

The first series for consideration is the sum of the squares of the natural numbers. Repeatedly using the expansion for $(1 + n)^2 = 1^2 + 2n + n^2$, we get the following:

$$(1 + 1)^2 = 1^2 + 2 \cdot 1 + 1^2$$

$$(1 + 2)^2 = 1^2 + 2 \cdot 2 + 2^2$$

$$(1 + 3)^2 = 1^2 + 2 \cdot 3 + 3^2$$

$$(1 + 4)^2 = 1^2 + 2 \cdot 4 + 4^2$$

$$(1 + 5)^2 = 1^2 + 2 \cdot 5 + 5^2$$

$$\vdots \quad \vdots$$

$$(1 + n)^2 = 1^2 + 2n + n^2.$$

Sum both sides of the above list of equations to get

$$2^2 + 3^2 + 4^2 + 5^2 + \cdots + (n + 1)^2$$
$$= n + 2(1 + 2 + 3 + 4 + 5 + \cdots + n) + 1^2 + 2^2 + 3^2 + 4^2 + 5^2 + \cdots + n^2,$$

or $(n + 1)^2 = (n + 1) + 2(1 + 2 + 3 + 4 + 5 + \cdots + n)$. Therefore,

$$1 + 2 + 3 + 4 + 5 + \cdots + n = \frac{(n+1)^2 - (n+1)}{2} = \frac{n(n+1)}{2}.$$

Let us now find the sum of the squares of the natural numbers: $1^2 + 2^2 + 3^2 + 4^2 + 5^2 + \cdots + n^2$.

First, let's try to use a cubic expansion, where the binomial theorem would be useful.

$$(1+1)^3 = 1^3 + 3 \cdot 1 + 3 \cdot 1^2 + 1^3$$

$$(1+2)^3 = 1^3 + 3 \cdot 2 + 3 \cdot 2^2 + 2^3$$

$$(1+3)^3 = 1^3 + 3 \cdot 3 + 3 \cdot 3^2 + 3^3$$

$$(1+4)^3 = 1^3 + 3 \cdot 4 + 3 \cdot 4^2 + 4^3$$

$$(1+5)^3 = 1^3 + 3 \cdot 5 + 3 \cdot 5^2 + 5^3$$

$$\vdots \qquad \vdots$$

$$(1+n)^3 = 1^3 + 3n + 3n^2 + n^3.$$

Sum both sides of the above equations to get

$$2^3 + 3^3 + 4^3 + 5^3 + \cdots + (n+1)^3$$
$$= n + 3(1 + 2 + 3 + 4 + 5 + \cdots + n) + 3(1^2 + 2^2 + 3^2 + 4^2 + 5^2 + \cdots + n^2)$$
$$+ (1^3 + 2^3 + 3^3 + 4^3 + 5^3 + \cdots + n^3). \text{ Or } (n+1)^3$$
$$= n + 3(1 + 2 + 3 + 4 + 5 + \cdots + n) + 3(1^2 + 2^2 + 3^2 + 4^2 + 5^2 + \cdots + n^2) + 1.$$

Substitute $1 + 2 + 3 + 4 + 5 + \cdots + n = \frac{n(n+1)}{2}$. We then have

$$1^2 + 2^2 + 3^2 + \cdots n^2 = \frac{(n+1)^3 - (n+1)}{3} - \frac{n(n+1)}{2}$$
$$= \frac{(n+2)n(n+1)}{3} - \frac{n(n+1)}{2} = \frac{n(n+1)(2n+1)}{6}.$$

We are now ready to find the sum of the cubes of the natural numbers: $1^3 + 2^3 + 3^3 + 4^3 + 5^3 + \cdots + n^3$.

Using a similar method as above, we expand $(1 + n)^4$ to find this sum:

$$(1+1)^4 = 1^4 + 4 \cdot 1 + 6 \cdot 1^2 + 4 \cdot 1^3 + 1^4$$
$$(1+2)^4 = 1^4 + 4 \cdot 2 + 6 \cdot 2^2 + 4 \cdot 2^3 + 2^4$$
$$(1+3)^4 = 1^4 + 4 \cdot 3 + 6 \cdot 3^2 + 4 \cdot 3^3 + 3^4$$
$$(1+4)^4 = 1^4 + 4 \cdot 4 + 6 \cdot 4^2 + 4 \cdot 4^3 + 4^4$$
$$(1+5)^4 = 1^4 + 4 \cdot 5 + 6 \cdot 5^2 + 4 \cdot 5^3 + 5^4$$
$$\vdots \qquad \vdots$$
$$(1+n)^4 = 1^4 + 4n + 6n^2 + 4n^3 + n^4.$$

Sum both sides of above equations to get

$$1^4 + 2^4 + 3^4 + 4^4 + 5^4 + \cdots + (n+1)^4$$
$$= n + 4 \cdot (1 + 2 + 3 + 4 + 5 + \cdots + n) + 6 \cdot (1^2 + 2^2 + 3^2$$
$$+ 4^2 + 5^2 + \cdots + n^2) + 4 \cdot (1^3 + 2^3 + 3^3 + 4^3 + 5^3 + \cdots + n^3)$$
$$+ 1^4 + 2^4 + 3^4 + 4^4 + 5^4 + \cdots + n^4. \text{ Or } (n+1)^4$$
$$= n + 1 + 4(1 + 2 + 3 + 4 + 5 + \cdots + n)$$
$$+ 6(1^2 + 2^2 + 3^2 + 4^2 + 5^2 + \cdots + n^2)$$
$$+ 4(1^3 + 2^3 + 3^3 + 4^3 + 5^3 + \cdots + n^3).$$

We substitute

$$1 + 2 + 3 + 4 + 5 + \cdots + n = \frac{n(n+1)}{2},$$

and

$$1^2 + 2^2 + 3^2 + \cdots + n^2 = \frac{n(n+1)(2n+1)}{6}$$

into the above equation to get

$$(n+1)^4 = (n+1) + 2n(n+1) + n(n+1)(2n+1)$$
$$+ 4(1^3 + 2^3 + 3^3 + 4^3 + 5^3 + \cdots + n^3).$$

Therefore, $1^3 + 2^3 + 3^3 + 4^3 + 5^3 + \cdots + n^3 = \frac{n^2(n+1)^2}{4}$.

Using a similar method, we can find the formula for the sum of a power series in the form $1^k + 2^k + 3^k + 4^k + 5^k + \cdots + n^k$, where k and n are integers.

Introducing Compound Interest using "The Rule of 72"

Compound interest is usually presented in textbooks through a question about how money will grow at a certain interest rate over a specified period of time. As a "real world" problem, it has all the necessary motivational attributes of a good introduction to the topic. However, it might be nice to try to use the "Rule of 72" to generate some interest in the compound interest formula.

Let us first introduce the famous "Rule of 72." It states that *money will double in $\frac{72}{r}$ years when it is invested at an annual compounded interest rate of r%*. For example, if we invest money at an 8% compounded annual interest rate, it will double its value in $\frac{72}{8} = 9$ years.

To investigate why or if this really works, we consider the compound interest formula $A = P\left(1 + \frac{r}{100}\right)^n$, where P is the principal invested for n interest periods at r% annually and A is the resulting amount of money.

We need to investigate what happens when $A = 2P$. The above equation then becomes

$$2 = \left(1 + \frac{r}{100}\right)^n. \tag{1}$$

It then follows that

$$n = \frac{\log 2}{\log\left(1 + \dfrac{r}{100}\right)}. \tag{2}$$

The table of values from the above equation is shown in Figure 5-3.

r	n	nr
1	69.66071689	69.66071689
3	23.44977225	70.34931675
5	14.20669908	71.03349541
7	10.24476835	71.71337846
9	8.043231727	72.38908554
11	6.641884618	73.0607308
13	5.671417169	73.72842319
15	4.959484455	74.39226682

Figure 5-3

Using Triangular Numbers to Generate Interesting Relationships

Triangular numbers are numbers that can be represented as points arranged in the form of an equilateral triangle, as shown in Figure 5-4.

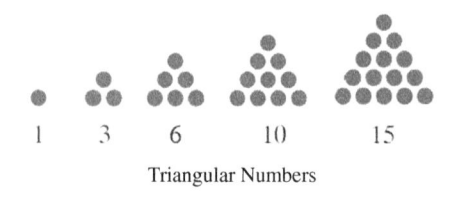

Triangular Numbers

Figure 5-4

We shall use the notation T_n to represent the nth triangular number. We notice that $T_1 = 1$, $T_2 = 1 + 2 = 3$, $T_3 = 1 + 2 + 3 = 6$, $T_4 = 1 + 2 + 3 + 4 = 10$, and $T_n = 1 + 2 + 3 + 4 + 5 + \cdots + n$.

Through some experimentation, we can see that

$$T_0 = 0,$$
$$T_0 + T_1 = 0 + 1 = 1,$$
$$T_1 + T_2 = 1 + 3 = 4,$$
$$T_2 + T_3 = 3 + 6 = 9,$$

and

$$T_3 + T_4 = 6 + 10 = 16,$$

which can be generalized to read

$$T_{n-1} + T_n = n^2. \tag{1}$$

Similar experimentation will yield the following:

$$T_0 = 0$$
$$T_1 - T_0 = 1$$
$$T_2 - T_1 = 2$$
$$T_3 - T_2 = 3 \tag{2}$$
$$T_4 - T_3 = 4,$$
$$\vdots$$
$$T_n - T_{n-1} = n.$$

When we add equations (1) and (2), we get a familiar result:

$$2T_n = n^2 + n = n(n+1)$$
$$T_n = \frac{1}{2}n(n+1),$$

which is the sum of the first n natural numbers.

Now consider the product of equations (1) and (2):

$$(T_{n-1} + T_n)(T_n - T_{n-1}) = n^2 \cdot n = n^3$$
$$T_n^2 - T_{n-1}^2 = n^3.$$

Each of the above can be justified inductively, but for the purposes of introducing some order in the number system, this treatment suffices. One ought to extend the current representation of powers of n as we have above. So far, we have

$$T_n - T_{n-1} = n$$
$$T_n + T_{n-1} = n^2$$
$$T_n^2 - T_{n-1}^2 = n^3.$$

Further representations are left as a challenge for the motivated reader.

Comparing Measures of Central Tendency

In mathematics and in statistics we frequently use measures of central tendency—that is, we use various means such as the *arithmetic mean* (in common terms: the average), the *harmonic mean*, and the *geometric mean*. Our knowledge of them dates back to ancient times. As a matter of fact, the historian Iamblichos of Chalkis (ca. 250–330 CE) reported that Pythagoras, after a visit to Mesopotamia, brought back to his followers a knowledge of these three measures of central tendency. This may be one reason why today they are often referred to as *Pythagorean means.* We tend to use these means in statistical analyses, but there are some rather enlightening revelations when we inspect them and compare them geometrically.

Let's begin by introducing these three means as follows for two values a and b:

- The *arithmetic mean* is $AM(a, b) = a \,Ⓐ\, b = \frac{a+b}{2}$.
- The *harmonic mean* is $HM(a, b) = a \,Ⓗ\, b = \frac{2}{\frac{1}{a}+\frac{1}{b}} = \frac{2ab}{a+b}$.
- The *geometric mean* is $GM(a, b) = a \,Ⓖ\, b = \sqrt{a \cdot b}$.[1]

---

[1] The arithmetic mean is defined for when a and b are all real numbers; for the harmonic mean, a and b are defined for all real numbers but not when $a + b = 0$; and the geometric mean is defined for when $a > 0$ and $b > 0$.

For convenience throughout this discussion, we will use these representations as we discuss them in greater detail.

The Arithmetic Mean

Before comparing the relative magnitude of these measures of central tendency, or means, we ought to see what they actually represent. The arithmetic mean is simply the commonly-used "average" of the data being considered—that is, the sum of the data divided by the number of data items included in the sum. In a simple example, if we want to find the average—or arithmetic mean—between the two values of 30 and 60, we take their sum, 90, and divide it by 2 to get 45.

We can also see the arithmetic mean as taking us to the middle of an arithmetic sequence—that is, a sequence with a common difference between terms, such as 2, 4, 6, 8, 10. To get the arithmetic mean, we take the sum divided by the number of numbers being averaged. Here, we get $\frac{2+4+6+8+10}{5} = \frac{30}{5} = 6$, which happens to be the middle number in the sequence.

The Harmonic Mean

If we take an arithmetic sequence such as 1, 2, 3, 4, 5, and take the reciprocals, we have a harmonic sequence: $1, \frac{1}{2}, \frac{1}{3}, \frac{1}{4}, \frac{1}{5}$. We can tie the harmonic mean to the arithmetic mean by indicating that the harmonic mean is the reciprocal of the arithmetic mean of the reciprocals of the numbers. To get the harmonic mean of a given sequence 1, 2, 3, 4, 5, we first find the arithmetic mean of the reciprocals of the sequence, that is, $1 + \frac{1}{2} + \frac{1}{3} + \frac{1}{4} + \frac{1}{5} = \frac{60+30+20+15+12}{60} = \frac{137}{60}$. We then take the reciprocal of this value to get the harmonic mean, $\frac{60}{137} \approx .438$.

The harmonic mean is particularly useful for finding the average of rates over a common base, as demonstrated in the previous section. For example, consider once again the problem of finding the average speed of a round trip journey. Suppose you were going at a rate of 30 mph and returning over the same route (the base) at 60 mph. As mentioned earlier in Arithmetic Novelty 65, one might be tempted to simply find the arithmetic mean, $\frac{30+60}{2} = 45$. This would be incorrect, as it is unfair to give equal value to the 30-mph trip and the 60-mph trip when the former took twice as long as the latter. Instead, we invoke the harmonic mean, which would require us to get the reciprocal of the arithmetic mean of the reciprocals of the two numbers. That is, for the harmonic mean of 30 and 60, we get

$$\frac{1}{\dfrac{\dfrac{1}{30}+\dfrac{1}{60}}{2}} = \frac{2}{\dfrac{1}{30}+\dfrac{1}{60}} = \frac{2}{\dfrac{3}{60}} = \frac{120}{3} = 40.$$

This could, of course, be more simply done by using the formula

$$\frac{1}{\dfrac{\dfrac{1}{a}+\dfrac{1}{b}}{2}} = \frac{2ab}{a+b}.$$

The Geometric Mean

The geometric mean gets its name from a simple interpretation of geometry. A rather common application of the geometric mean is obtained when we consider the altitude to the hypotenuse of a right triangle. In Figure 5-5, we have *CD* as the altitude to the hypotenuse of right triangle *ABC*.

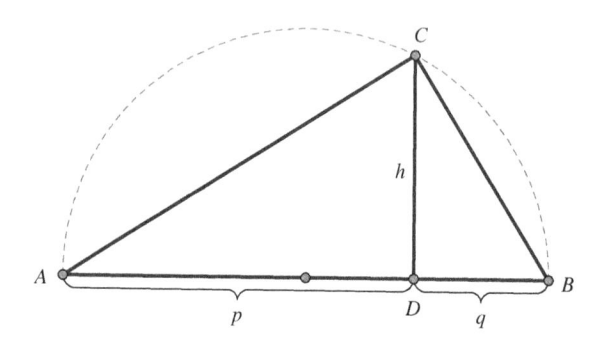

Figure 5-5

From the triangle similarity $\triangle ADC \sim \triangle CDB$, we get $\frac{AD}{CD} = \frac{CD}{BD}$, or $\frac{p}{h} = \frac{h}{q}$. This then gives us $h = \sqrt{pq}$, which has h as the geometric mean between p and q. The geometric mean is also the "middle" of a geometric sequence.[2] Take, for example, the geometric sequence 2, 4, 8, 16, 32. To get the geometric mean of these five numbers, we find the fifth root of their product: $\sqrt[5]{2 \cdot 4 \cdot 8 \cdot 16 \cdot 32} = \sqrt[5]{32,768} = 8$, which is the middle number.

Comparing the Three Means

Before we present some unusual geometric methods for comparing the magnitude of these three means, we shall now show how these three means may be compared in size using simple algebra.

Comparing the arithmetic and geometric mean of the two non-negative numbers a and b:

$$(a - b)^2 \geq 0$$

$$a^2 - 2ab + b^2 \geq 0.$$

Add $4ab$ to both sides:

$$a^2 + 2ab + b^2 \geq 4ab$$

$$(a + b)^2 \geq 4ab.$$

Take the positive square root of both sides:

$$a + b \geq 2\sqrt{ab}$$

$$\text{or } \frac{a+b}{2} \geq \sqrt{ab}.$$

This implies that the *arithmetic mean* is greater than, or equal to, the *geometric mean*. (Equality is true only if $a = b$.)

---

[2] A geometric sequence is one with a common factor between consecutive terms.

Beginning as we did above and then continuing along as shown below, we get our next desired result: a comparison of the geometric mean and the harmonic mean.

$$(a-b)^2 \geq 0$$

$$a^2 - 2ab + b^2 \geq 0.$$

Add $4ab$ to both sides:

$$a^2 + 2ab + b^2 \geq 4ab$$

$$(a+b)^2 \geq 4ab.$$

Multiply both sides by ab:

$$ab(a+b)^2 \geq (4ab)(ab)$$

$$ab \geq \frac{4a^2b^2}{(a+b)^2}.$$

Take the positive square root of both sides:

$$\sqrt{ab} \geq \frac{2ab}{a+b}.$$

This tells us that the *geometric mean* is greater than or equal to the *harmonic mean*. (Here, equality holds whenever one of these numbers is zero or if $a = b$.)

We can then conclude that *arithmetic mean $\geq$ geometric mean $\geq$ harmonic mean*.

Comparing the Three Means Geometrically Using a Rectangle

The comparison of the three means in terms of their relative size was known to the ancient Greeks, as we find in the writings of Pappus of Alexandria (ca. 250–350 CE). We will now embark on a geometric journey to consider various ways that the relative sizes of these means can be compared using simple geometric relationships.

In Figure 5-6, we have a right triangle with an altitude drawn to the hypotenuse. The hypotenuse is partitioned by the altitude into two segments of lengths a and b, and where $a < b$. Figure 5-6 shows the line segments that can represent the three means, so that we can visually see their relative sizes. That is, $a \, \text{Ⓗ} \, b < a \, \text{Ⓖ} \, b < a \, \text{Ⓐ} \, b$.

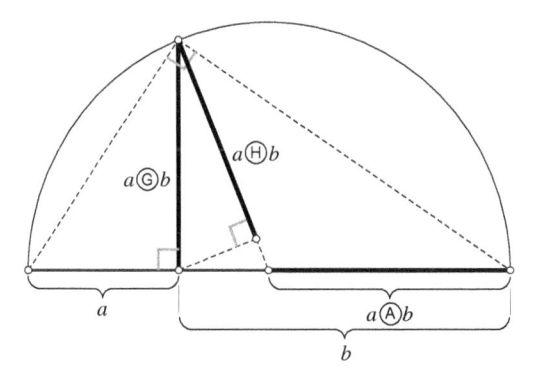

Figure 5-6

To justify our visual demonstration, we begin by considering Figure 5-7, where we notice that CE is a leg of right triangle CED; therefore, $CE < CD$. Since the radius of the circumscribed circle of triangle ABC is longer than the altitude to the hypotenuse of the triangle ABC, we have $CD < MB$. Combining these inequalities gives us $CE < CD < MB$. Our task now is to show that these three segments,

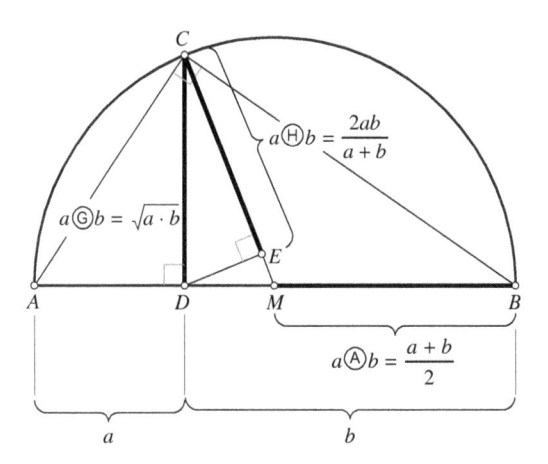

Figure 5-7

whose relative lengths we now have established, actually represent the three means of a and b as we earlier stated.

First, we know that radii $MA = MB = MC = \frac{a+b}{2}$, which is the arithmetic mean between a and b. To find the geometric mean of a and b we begin with the altitude CD to the hypotenuse of right triangle ABC, which partitions the right triangle into two similar triangles ($\triangle ADC \sim \triangle CDB$). Therefore, $\frac{a}{CD} = \frac{CD}{b}$, which leads to $CD^2 = a \cdot b$; thus, $CD = \sqrt{a \cdot b}$, which is the geometric mean of a and b. From the similar triangles ($\triangle CDM \sim \triangle ECD$) located in right triangle CDM, we can get $CD^2 = MC \cdot CE$, yielding $CE = \frac{CD^2}{CM} = \frac{a \cdot b}{\frac{a+b}{2}} = \frac{2ab}{a+b}$, which is the harmonic mean.

Having now justified how the line segments, which we size-ordered as $CE < CD < MB$, can represent the various means of a and b, we have thus shown geometrically that $\frac{2ab}{a+b} \leq \sqrt{a \cdot b} \leq \frac{a+b}{2}$, or that *arithmetic mean* $\geq$ *geometric mean* $\geq$ *harmonic mean*.

The Magic of Algebra

In today's world, complicated calculations are easily disposed of using a calculator. However, it could be entertaining to see how, through algebraic manipulation, we can make a very complicated calculation practically trivial.

Consider the task of finding the value of $\sqrt{1{,}999 \cdot 2{,}000 \cdot 2{,}001 \cdot 2{,}002 + 1}$. Using a calculator, we can find that this cumbersome expression is equal to 4,001,999. However, we can generalize this expression and make it much simpler to solve. Since the numbers being multiplied are consecutive, we might see if we can use this pattern to our advantage. If we begin by letting $n = 2{,}000$, we can express the other numbers under the radical sign in this way: $(n-1) \cdot n \cdot (n+1) \cdot (n+2) + 1$. Now for some algebraic gymnastics. By multiplying the terms of this long algebraic expression, and then adding 1, we get $(n-1) \cdot n \cdot (n+1) \cdot (n+2) + 1 = n^4 + 2n^3 - n^2 - 2n + 1$. Now, we rearrange and dismantle these terms to suit our plan to get a workable expression: $n^4 + 2n^3 - n^2 - 2n + 1 = n^4 + n^3 - n^2 + n^3 + n^2 - n - n^2 - n + 1$. This allows us to form the following product of two trinomials: $n^4 + n^3 - n^2 + n^3 + n^2 - n - n^2 - n + 1 = (n^2 + n - 1) \cdot (n^2 + n - 1) = (n^2 + n - 1)^2$. By replacing the original term under the radical sign with its equivalent established above, we can simplify the expression under the radical—a perfect square—which allows us to remove the radical sign: $\sqrt{(n-1)\cdot n\cdot(n+1)\cdot(n+2)+1} = \sqrt{(n^2+n-1)^2} = |n^2+n-1|$. Because we are working with natural numbers, we can conclude that $\sqrt{(n-1)\cdot n\cdot(n+1)\cdot(n+2)+1} = \sqrt{(n^2+n-1)^2} = n^2 + n - 1$. Therefore, when $n = 2{,}000$, we get $\sqrt{1{,}999\cdot 2{,}000\cdot 2{,}001\cdot 2{,}002+1} = 2{,}000^2 + 2{,}000 - 1 = 4{,}000{,}000 + 2{,}000 - 1 = 4{,}001{,}999$, which conforms to the result obtained by using a calculator.

Perfect Numbers

You may be wondering: what makes a number perfect? Mathematicians define a number to be a *perfect number* when the sum of its factors (excluding the number itself) is equal to the number. For example, the smallest perfect number is 6, because the sum of its factors (of its proper positive divisors) is $1 + 2 + 3 = 6$. The next perfect number is 28, since the sum of its factors is $1 + 2 + 7 + 4 + 14 = 28$. Perfect numbers have fascinated mathematicians for centuries. The ancient Greeks knew of the first four perfect numbers, and Euclid (ca. 300 BCE) established a formula for generating these perfect numbers. These first two perfect numbers were esteemed by ancient biblical scholars, as the biblical creation was accomplished in six days and the lunar month has 28 days. The next two perfect numbers (496 and 8,128) are attributed to Nicomachus (ca. 100 CE). Yet it was not until the eighteenth century that the Swiss mathematician Leonhard Euler (1707–1783) proved that Euclid's formula would generate all even perfect numbers.

The formula for generating even perfect numbers is $2^{n-1}(2^n - 1)$, with the condition that $2^n - 1$ is a prime number. The value of $2^n - 1$ is referred to as a Mersenne prime number, M_p,[3] named after the French mathematician Marin Mersenne (1588–1648), who studied perfect numbers. For one thing, if n is a composite number, then $2^n - 1$ will surely be composite, thereby making $n \neq 1$ to generate a perfect

---

[3] $M_p = 2^p - 1$ and p is prime.

number from Euclid's formula. One can demonstrate this with some very elementary algebra, as we will show here.

Suppose n is an even composite number, say $2x$. the expression $2^n - 1$ then becomes the difference of two squares, which is always factorable, and is, therefore, a composite number: $2^{2x} - 1 = (2^x - 1)(2^x + 1)$. If n is an odd composite number, as in the expression $2^{pq} - 1$, it becomes a factorable term as

$$2^{pq} - 1 = (2^p)^q - 1 = (2^p - 1)((2^p)^{q-1} + (2^p)^{q-2} + (2^p)^{q-3} + \cdots + (2^2) + 1).$$

This is not to say that whenever n is prime, $2^n - 1$ will also be prime. For example, when $n = 11$, we have $2^{11} - 1 = 2,048 - 1 = 2,047 = 23 \cdot 89$, which is not prime. Therefore, we have to be careful to make sure that n is a prime number and that it also makes $2^n - 1$ a prime.

There have been some bumps in the road that mathematicians encountered as they pursued the search for further perfect numbers. For example, Mersenne corrected the list of 24 perfect numbers published by the sixteenth century mathematician Peter Bungus in his book *Numerorum Mysteria* (1591), stating that only eight of these were correct (i.e., the first eight listed in the table in Figure 5-8). However, Mersenne offered three more numbers to this list of perfect numbers (namely, those where n in Euclid's formula had values 67, 127, and 257). By 1947, it was proved that only $n = 127$ was correct; at that time, two more perfect numbers with n values of 89 and 107 were added to the list of perfect numbers.

Mathematicians have always been fascinated with perfect numbers, and consequently are always searching for further members of this set. As of the publication of this book, there are 51 known perfect numbers, as shown in the table in Figure 5-8.

Perfect numbers possess lots of curious characteristics. For example, notice that they all end in either 28 or 6, and that those digits are preceded by an odd digit. Mathematicians have been hunting for odd perfect numbers, but as we can see, our list consists only of even perfect numbers. To date, they say, with confidence, that there are no odd perfect numbers less than 10^{1500}.

Rank	n	Perfect number	Digits	Year discovered
1	2	6	1	Greeks
2	3	28	2	Greeks
3	5	496	3	Greeks
4	7	8128	4	Greeks
5	13	33550336	8	1456
6	17	8589869056	10	1588
7	19	137438691328	12	1588
8	31	2305843008139952128	19	1772
9	61	2658455991569831744654692 615953842176	37	1883
10	89	19156194260823610729479 33 78084303638130997 321548169216	54	1911
11	107	131640364 ... 783728128	65	1914
12	127	144740111 ... 199152128	77	1876
13	521	235627234 ... 555646976	314	1952
14	607	141053783 ... 537328128	366	1952
15	1279	541625262 ... 984291328	770	1952
16	2203	108925835 ... 453782528	1327	1952
17	2281	994970543 ... 139915776	1373	1952
18	3217	335708321 ... 628525056	1937	1957
19	4253	182017490 ... 133377536	2561	1961
20	4423	407672717 ... 912534528	2663	1961
21	9689	114347317 ... 429577216	5834	1963
22	9941	598885496 ... 073496576	5985	1963
23	11213	395961321 ... 691086336	6751	1963
24	19937	931144559 ... 271942656	12003	1971
25	21701	100656497 ... 141605376	13066	1978
26	23209	811537765 ... 941666816	13973	1979
27	44497	365093519 ... 031827456	26790	1979

Figure 5-8

Rank	n	Perfect number	Digits	Year discovered
28	86243	144145836 … 360406528	51924	1982
29	110503	136204582 … 603862528	66530	1988
30	132049	131451295 … 774550016	79502	1983
31	216091	278327459 … 840880128	130100	1985
32	756839	151616570 … 565731328	455663	1992
33	859433	838488226 … 416167936	517430	1994
34	1257787	849732889 … 118704128	757263	1996
35	1398269	331882354 … 723375616	841842	1996
36	2976221	194276425 … 174462976	1791864	1997
37	3021377	811686848 … 022457856	1819050	1998
38	6972593	955176030 … 123572736	4197919	1999
39	13466917	427764159 … 863021056	8107892	2001
40	20996011	793508909 … 206896128	12640858	2003
41	24036583	448233026 … 572950528	14471465	2004
42	25964951	746209841 … 791088128	15632458	2005
43	30402457	497437765 … 164704256	18304103	2005
44	32582657	775946855 … 577120256	19616714	2006
45	37156667	204534225 … 074480128	22370543	2008
46	42643801	144285057 … 377253376	25674127	2009
47	43112609	500767156 … 145378816	25956377	2008
48	57885161	169296395 … 270130176	34850340	2013
49	74207281	109200 … 301056	44677235	2016
50	77232917	110847 … 207936	46498850	2017
51	82589933	110847 … 207936	49724095	2018

Figure 5-8   (*Continued*)

There are many unusual characteristics of perfect numbers beyond those that define them—namely, that the sum of their divisors equals the number itself. Here are now some curious characteristics of perfect numbers of the form $2^{n-1}(2^n - 1)$. First, they are the sum of the first consecutive natural numbers:

$$6 = 1 + 2 + 3$$
$$28 = 1 + 2 + 3 + 4 + 5 + 6 + 7$$
$$496 = 1 + 2 + 3 + 4 + 5 + 6 + 7 + 8 + 9 + 10 + 11 + 12 + 13$$
$$+ 14 + \cdots + 29 + 30 + 31$$
$$8,128 = 1 + 2 + 3 + 4 + 5 + 6 + 7 + 8 + 9 + 10 + 11 + 12 + 13$$
$$+ 14 + \cdots + 125 + 126 + 127$$
$$33,550,336 = 1 + 2 + 3 + 4 + 5 + 6 + 7 + 8 + 9 + 10 + 11 + 12$$
$$+ 13 + \cdots + 8,189 + 8,190 + 8,191,$$

and so on.

If this isn't enough, we also notice that with the exception of the first perfect number, 6, the remaining perfect numbers are equal to the sum of odd cubes as seen below:

$$28 = 1^3 + 3^3$$
$$496 = 1^3 + 3^3 + 5^3 + 7^3$$
$$8,128 = 1^3 + 3^3 + 5^3 + 7^3 + 9^3 + 11^3 + 13^3 + 15^3$$
$$33,550,336 = 1^3 + 3^3 + 5^3 + \cdots + 123^3 + 125^3 + 127^3.$$

How can it be that perfect numbers are equal to the sum of the cubes of a sequence of odd numbers? This can be easily justified through some elementary algebra. We recall that the perfect number must take the form of $2^{n-1}(2^n - 1)$, where $2^n - 1$ is a prime number. We will take a moment here to show how we can justify that each perfect number is the sum of the first 2^k odd numbers, where $k = \frac{1}{2}(n-1)$, except for $n = 2$.

We should recall that the following are true:

$$S_1 = 1 + 2 + 3 + 4 + \cdots + q = \frac{q(q+1)}{2}$$

$$S_2 = 1^2 + 2^2 + 3^2 + 4^2 + \cdots + q^2 = \frac{q(q+1)(2q+1)}{6}$$

$$S_3 = 1^3 + 2^3 + 3^3 + 4^3 + \cdots + q^3 = \frac{q^2(q+1)^2}{4} = S_1^2.$$

Let us now look at the sum of the cubes of the odd numbers. We can write that as follows:

$$S = 1^3 + 3^3 + 5^3 + 7^3 + \cdots + (2q-1)^3 = \sum_{i=1}^{q} (2q-1)^3.$$

With some algebraic manipulation,[4] we can show that this is equal to the following:

$$S = \sum_{i=1}^{q} (2q-1)^3 = \sum_{i=1}^{q} \left(8q^3 - 12q^2 + 6q - 1\right)$$

$$= 8 \cdot \frac{q^2(q+1)^2}{4} - 12 \cdot \frac{q(q+1)(2q+1)}{6} + 6 \cdot \frac{q(q+1)}{2} - q$$

$$= q^2(2q^2 - 1).$$

Notice how we inserted the first three formulas that we generated above into this last equation.

If we now let $q = 2^k$, then $S = 2^{2k}(2^{2k+1} - 1)$, and we notice that this would generate the perfect numbers $2^{n-1}(2^n - 1)$ when $n = 2k + 1$, which are the odd numbers. This, then, shows how we go from the general formula for the perfect numbers to the sum of the cubes of the odd numbers. Although this may involve more sophisticated elementary algebra, we provided it here to show how we can justify some of these mathematical curiosities.

All perfect numbers must have an even number of divisors. If we take the reciprocals of the divisors of any perfect number (now including the number itself), their sum will always be equal to 2. This can be seen from the first few perfect numbers as follows:

$$\frac{1}{1} + \frac{1}{2} + \frac{1}{3} + \frac{1}{6} = 2$$

$$\frac{1}{1} + \frac{1}{2} + \frac{1}{4} + \frac{1}{7} + \frac{1}{14} + \frac{1}{28} = 2$$

$$\frac{1}{1} + \frac{1}{2} + \frac{1}{4} + \frac{1}{8} + \frac{1}{16} + \frac{1}{31} + \frac{1}{62} + \frac{1}{124} + \frac{1}{248} + \frac{1}{496} = 2.$$

---

[4] Working backwards.

Algebra Reveals a Curiosity

We know from algebra that squaring a binomial, such as $x + 1$, gives us $(x + 1)^2 = x^2 + 2x + 1 = x^2 + (2x + 1)$. This tells us that to get from one square number, say x^2, to the next square number, $(x + 1)^2$, we simply have to add twice x plus 1 (that is, $2x + 1$), since $(x + 1)^2 = x^2 + (2x + 1)$. To see how this works, we can take the square number $4^2 = 16$, find $2x + 1 = 4 \cdot 2 + 1 = 9$, and then add this to the square to get the next square number: $16 + 9 = 25 = 5^2$.

Thus, the difference between consecutive squares is twice the square root of the first square plus 1. We can show this with the following example. The difference between the squares 64 and 81 can be found by taking twice the square root of 64 and adding 1: $2 \cdot 8 + 1 = 17$, which is equal to $81 - 64$. This gives us a nice procedure of finding the difference between any two square numbers.

The Mystical Number 1089

Some numbers are especially fascinating. One such number is 1,089, which has a variety of unusual characteristics. For example, take the reciprocal of this number (1,089) to get the following: $\frac{1}{1089} = .\overline{000918273645546372819}$. With the exception of the first three zeros and the last 1, we have a palindromic number: 918,273,645,546,372,819, since it reads the same in both directions. Furthermore, if we multiply 1,089 by 5, we get the palindromic number 5,445. If we multiply 1,089 by 9, we get 9,801, whose digits are in the reverse order of the original number. The only other number of four or fewer digits whose multiple is the reverse of the original number is 2178, since $2178 \cdot 4 = 8,712$.

Let us now multiply by 9 of some modifications of 1,089, say 10,989, 109,989, 1,099,989, 10,999,989, and so on, and then marvel at the results.

$$10,989 \cdot 9 = 98,901$$
$$109,989 \cdot 9 = 989,901$$
$$1,099,989 \cdot 9 = 9,899,901$$
$$10,999,989 \cdot 9 = 98,999,901.$$

and so on.

This algebraic justification enables us to inspect the arithmetic process, regardless of the original number, and thereby generalize it to all such numbers.

Returning to 1,089, here's another neat little trick. Select any three-digit number whose units digit and hundreds digits are not the same, and then reverse that number. Now subtract the two numbers you have (the larger minus the smaller). Once again reverse the digits of this difference and add this new number to the difference. The result will *always* be 1,089. To see how this works, we will randomly select a three-digit number, say 732. We now subtract $732 - 237 = 495$. Reversing the digits of 495 we get 594, and now we add these last two numbers: $495 + 594 = 1,089$. Yes, this will hold true for all such three-digit numbers! Amazing!

We can justify this cute little "trick" using elementary algebra. We shall represent the arbitrarily selected three-digit number $\overline{htu}$ as $100h + 10t + u$, where h represents the hundreds digit, t represents the tens digit, and u represents the units digit. The number with the digits reversed is then $100u + 10t + h$. We will let $h > u$, which would be the case either in the number you selected or the reverse of it. In the subtraction, $u - h < 0$; therefore, take 1 from the tens place (of the minuend), making the units place $10 + u$. Since the tens digits of the two numbers to be subtracted are equal, and 1 was taken from the tens digit of the minuend, then the value of this digit is $10(t - 1)$. The hundreds digit of the minuend is $h - 1$, because 1 was taken away to enable subtraction in the tens place, making the value of the tens digit $10(t - 1) + 100 = 10(t + 9)$.

When we do the first subtraction, we get

$$
\begin{array}{lll}
100(h-1) & +10(t+9) & +(u+10) \\
100u & +10t & +h \\
\hline
100(h-u-1) & +10(9) & +u-h+10
\end{array}.
$$

Reversing the digits of this difference

$$100(h - u - 1) + 10 \cdot 9 + (u - h + 10)$$

gives us

$$100(u - h + 10) + 10 \cdot 9 + (h - u - 1).$$

Adding these last two expressions gives us

$$100(h - u - 1) + 10 \cdot 9 + (u - h + 10) + 100(u - h + 10) + 10 \cdot 9 + (h - u - 1) = 1000 + 90 - 1 = \mathbf{1{,}089},$$

which is the constant surprise result!

Chapter 6

Neglected Topics in Geometry

Apart from the general education that is provided throughout the world, the United States offers a full year of geometry instruction, typically in the sophomore year of high school. This concentrated study of plane geometry is enhanced with some three-dimensional geometry, and everything that is presented must be proved (except for postulates and axioms) before it is accepted. This not only gives students a wonderful experience in an important field of mathematics, but it also provides a peek into how mathematicians function. However, there are still many aspects of geometry that are not explored because of the limited time available during the school year. This section provides a few topics that could easily be a part of the high school curriculum if there were more time to study geometry. Here you will see some of the geometric discoveries by famous people such as Napoleon Bonaparte and United States President James A. Garfield. Lots of geometric wonders await you.

The Pythagorean Theorem

We shall begin by asking the famous question: what do the following three men have in common: Pythagoras, Euclid, and James A. Garfield (1831–1881), the twentieth president of the United States? After some moments of perplexity, we shall relieve your frustration by telling you that all three fellows proved the Pythagorean theorem. The first two fellows should not be much of a surprise, but President Garfield? He wasn't a mathematician. He didn't even study mathematics. As a matter of fact, his only study of geometry, some 25 years before he published his proof of the Pythagorean theorem, was informal and alone.[1] While a member of the United States House of Representatives, Garfield, who enjoyed "playing" with elementary mathematics, stumbled upon a cute proof of this famous theorem, one that could easily be included in the high school curriculum, but somehow is not. It was subsequently published in the *New England Journal of Education* after being encouraged by two professors (Quimby and Parker) at Dartmouth College, where he went to give a lecture on March 7, 1876. The text begins with the following:

> "In a personal interview with General James A. Garfield, Member of Congress from Ohio, we were shown the following demonstration of the pons asinorum,[2] which he had hit upon in some mathematical amusements and discussions with other M.C.'s. We do not remember to have seen it before, and we think it something on which the members of both houses can unite without distinction of party."

Garfield's proof is actually quite simple, and therefore, can be considered "beautiful." Beauty is often found in simplicity! We begin the proof by placing two congruent right triangles ($\triangle ABE \cong \triangle DCE$) so that points B, C and E are collinear, as shown in Figure 6-1, whereupon a trapezoid is formed. Notice also that since $\angle AEB + \angle CED = 90°$, we get $\angle AEB = 90°$, making $\triangle AED$ a right triangle.

---

[1] In October 1851 he noted in his diary that "I have today commenced the study of geometry alone without class or teacher".

[2] This would appear to be a wrong reference, since we usually consider the proof that the base angles of an isosceles triangle are congruent as the pons asinorum, or "bridge of fools."

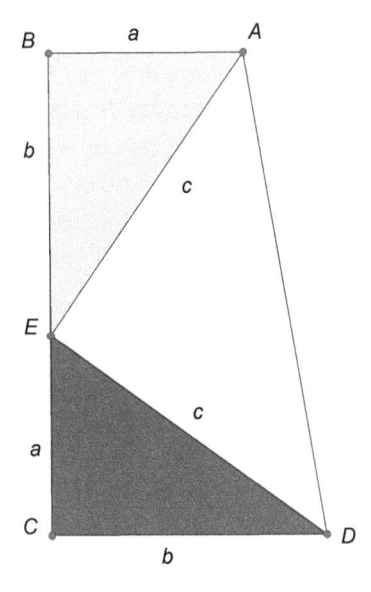

Figure 6-1

The area of the trapezoid $= \dfrac{1}{2}(\text{sum of bases}) \cdot (\text{altitude})$

$$= \dfrac{1}{2}(a+b)(a+b)$$

$$= \dfrac{1}{2}a^2 + ab + \dfrac{1}{2}b^2.$$

The sum of the areas of the three triangles, which is essentially the area of the trapezoid, is

$$= \dfrac{1}{2}ab + \dfrac{1}{2}ab + \dfrac{1}{2}c^2 = ab + \dfrac{1}{2}c^2.$$

We now equate the two expressions of the area of the trapezoid:

$$\dfrac{1}{2}a^2 + ab + \dfrac{1}{2}b^2 = ab + \dfrac{1}{2}c^2$$

$$\dfrac{1}{2}a^2 + \dfrac{1}{2}b^2 = \dfrac{1}{2}c^2,$$

which is the familiar $a^2 + b^2 = c^2$, known as the *Pythagorean theorem*.

There are over 400 proofs[3] of the Pythagorean theorem available today; many are ingenious, but some are a bit cumbersome. However, no proof can use trigonometry[4] because trigonometry depends (or is based) on the Pythagorean theorem, as $\cos^2\phi + \sin^2\phi = 1$. Thus, using trigonometry to prove the very theorem on which it depends would be circular reasoning. Have you been motivated to discover a new proof of this most famous theorem? Lots await you!

The Lunes and the Triangle

A lune is a crescent shape, like the waxing or waning moon, and is formed by two circular arcs. It is important to note as we progress in this topic that the area of a circle is typically not commensurate with the areas of rectilinear figures. A case in point is "squaring the circle," one of the "Three Famous Problems of Antiquity," which establishes that it is impossible to construct a square (with unmarked straight-edge and compasses) equal in area to a given circle. At first, this appears to be a case of incommensurability, where one cannot get a region formed by straight lines equal to one bounded by circular arcs. However, we shall provide a delightfully simple example where a circular area is, in fact, equal to the area of a triangle.

Let's first recall the Pythagorean theorem. It states that *the sum of the squares* of *the legs of a right triangle is equal to the square* of *the hypotenuse.* This can be restated as *the sum of the squares* on *the legs of a right triangle is equal to the square* on *the hypotenuse.* We can take this a step further by stating that *the sum of the* areas *of the squares on the legs of a right triangle is equal to the* area *of the square on the hypotenuse.*

As a matter of fact, we can easily show that the square can be replaced by any similar figures drawn on the sides of a right triangle so that *the sum of the* areas *of the* similar polygons *on the legs of a right triangle is equal to the* area *of the* similar polygon *on the hypotenuse.*

This can then be restated for the specific case of semicircles (which are, of course, similar) so that *the sum of the* areas *of the*

---

[3] A classic source for 370 proofs of the Pythagorean theorem is Elisha s. Loomis *The Pythagorean Proposition* (Reston, VA: NCTM, 1968).

[4] This years-long assumption has been put into question by a trigonometric proof presented by two high school students: Calcea Johnson and Ne'Kiya Jackson.

semicircles *on the legs of a right triangle is equal to the* area *of the* semicircle *on the hypotenuse.*

Thus, for Figure 6-2, the area of the semicircles relate as **Area Q + Area R = Area P.**

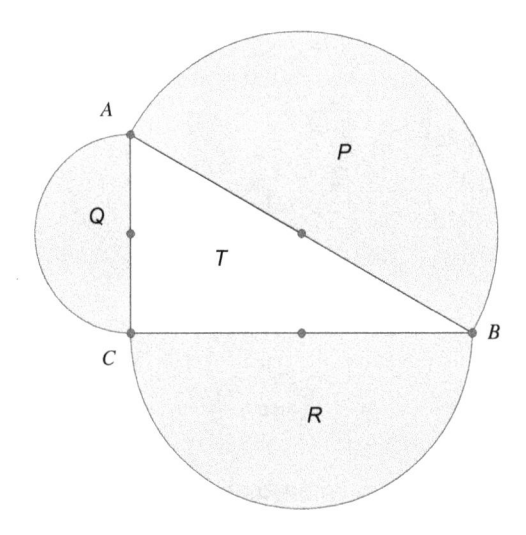

Figure 6-2

Suppose we now flip semicircle *P* over the rest of the figure by using *AB* as its axis. We would get a figure as shown in Figure 6-3.

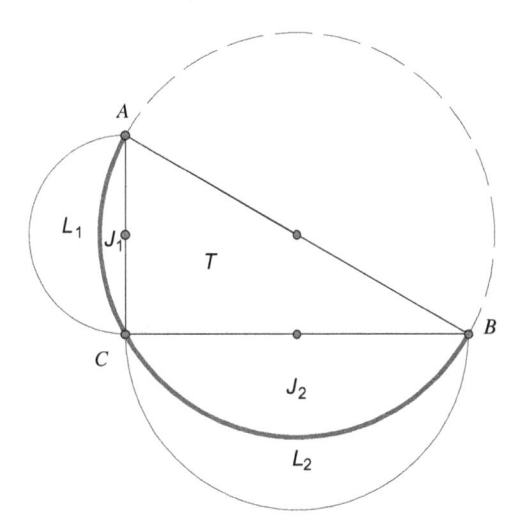

Figure 6-3

Let us now focus on the lunes, L_1 and L_2, formed by the two semi-circles, as shown in Figure 6-4.

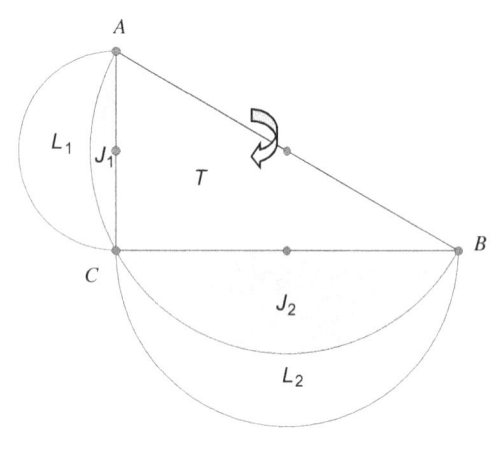

Figure 6-4

In Figure 6-2, we established that *Area P = Area Q + Area R.* Yet in Figure 6-4, that same relationship can be written as follows:

Area J_1 + Area J_2 + Area T = Area L_1 + Area J_1 + Area L_2 + Area J_2.

If we subtract *Area J_1 + Area J_2* from both sides, we get the astonishing result that *Area T = Area L_1 + Area L_2.* That is, we have the area of a rectilinear figure (the triangle) equal to the sum of the areas of non-rectilinear figures (the lunes).

Generating Pythagorean Triples

Pythagorean triples are numbers that satisfy the Pythagorean relationship. They are encountered in the high school geometry curriculum and are a normal part of any discussion of the Pythagorean theorem, but there is much more to be said about these interesting numbers. Other than the delight of having integers that fit the relationship $a^2 + b^2 = c^2$, special attention is rarely paid to the Pythagorean triples. Perhaps, if we consider some of the nice relationships that can be found through simple explorations, these triples would get the prominence that they deserve.

The most popular triple is (3, 4, 5), where $3^2 + 4^2 = 5^2$. This triple leads directly to others that are multiples of this triple, such as (6, 8, 10), (9, 12, 15), (30, 40, 50), etc. We should limit our discussion here to the primitive triples, those whose greatest common factor is 1. You may recall other primitive Pythagorean triples, such as: (5, 12, 13), (8, 15, 17), (7, 24, 25), (9, 40, 41), (12, 35, 37), etc. Let's consider some properties of these primitive Pythagorean triples:

- Two numbers of the triple are odd and one number is even.
- Some triples consist of numbers which differ by one.
- The product of the numbers of a triple is divisible by 60.
- One of the numbers of a triple is divisible by 5.

More properties can be found by considering the parametric equations that generate Pythagorean triples. Parametric equations are equations in which an original set of variables is expressed in terms of a second set of variables called the parameters. By assigning values to the parameters, we can obtain a set of values for our original variables. Although students rarely use parametric equations in their early work in algebra, they are often quite important in later study of mathematics and are worthwhile getting acquainted with early. In the case of the Pythagorean triples, the parametric equations are particularly useful. Here is a set of parametric equations that can be used to generate sets of Pythagorean triples:

$$a = u^2 - v^2$$

$$b = 2uv$$

$$c = u^2 + v^2,$$

where a and b are the lengths of the legs of the right triangle and c is the length of its hypotenuse, that is, $a^2 + b^2 = c^2$. By setting up a table of values (Figure 6-5) for these parametric equations, one can get an insight into some of the characteristics noted above.

u	v	$u^2 - v^2$	$2uv$	$u^2 + v^2$
2	1	3	4	5
3	2	5	12	13
4	1	15	8	17
4	3	7	24	25
5	4	9	40	41
3	1	8	6	10
5	2	21	20	29
6	1	35	12	37
6	5	11	60	61
7	2	45	28	53

Figure 6-5

The table in Figure 6-5 helps us make conjectures about the nature of special Pythagorean triples based on the values of u and v. For example, what must be true of u and v so that the triple is primitive? Or what must be true about u and v for two of the elements to differ by one? We can then ask further questions, and take our discussions in a variety of directions.

Menelaus' Theorem

The collinearity (that is, points lying on the same line) of three points is a topic that is rarely touched upon during high school geometry. However, a famous theorem proving collinearity is a very valuable tool in developing other interesting geometric relationships. In Figure 6-6, we have triangle ABC with three points, D, E, and F, on sides BC (extended), CA, and AB, respectively. Menelaus' theorem states that the three points D, E, and F are collinear if, and only if, the alternate segments they determine on the triangle sides have equal products. This can be more clearly expressed as an equation: $AF \cdot BD \cdot CE = AE \cdot BF \cdot CD$.

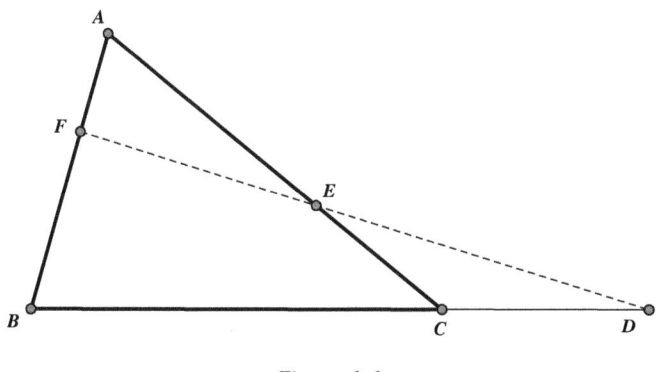

Figure 6-6

Proof: We will divide the proof of this theorem into several steps. First, we will assume that D, E, and F are collinear, and we will show that $AF \cdot BD \cdot CE = AE \cdot BF \cdot CD$. In Figure 6-7, we draw a line containing point C, parallel to AB, and intersecting DEF at point P.

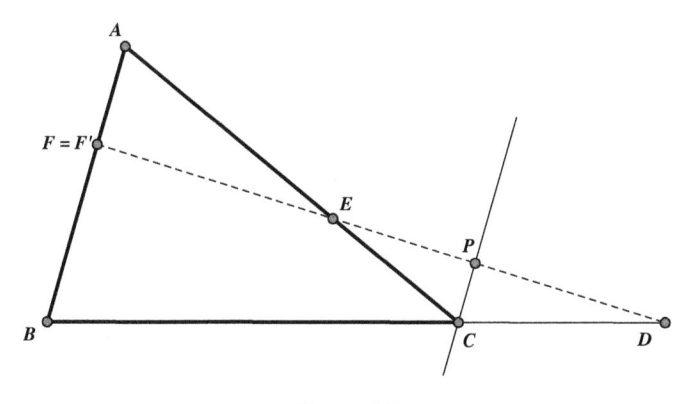

Figure 6-7

Since triangles PCD and FBD are similar, we have $\frac{CP}{BF} = \frac{CD}{BD}$, or $CP = \frac{BF \cdot CD}{BD}$. Furthermore, triangles PEC and FEA are also similar, so we then have $\frac{CP}{AF} = \frac{CE}{AE}$, or $CP = \frac{AF \cdot CE}{AE}$. This gives us $\frac{BF \cdot CD}{BD} = \frac{AF \cdot CE}{AE}$, which is equivalent to $AF \cdot BD \cdot CE = AE \cdot BF \cdot CD$, which proves the first part of Menelaus' theorem.

Next, we assume that $AF \cdot BD \cdot CE = AE \cdot BF \cdot CD$, and must show that points D, E, and F are collinear. We let the intersection point of AB

and DE be point F', shown in Figure 6-8. Then we have to prove $F' = F$. Since points D, E, and F' are collinear, we have just shown that $AF' \cdot BD \cdot CE = AE \cdot BF' \cdot CD$, which can also be written as

$$\frac{AE}{CE} \cdot \frac{BF'}{AF'} \cdot \frac{CD}{BD} = 1.$$

Since we are assuming that $AF \cdot BD \cdot CE = AE \cdot BF \cdot CD$, which we can also write as

$$\frac{AE}{CE} \cdot \frac{BF}{AF} \cdot \frac{CD}{BD} = 1,$$

we, therefore, have

$$\frac{BF'}{AF'} = \frac{BF}{AF}$$

and so we have shown that $F = F'$.

As a last step in our proof of Menelaus' theorem, we now consider the configuration illustrated in Figure 6-8.

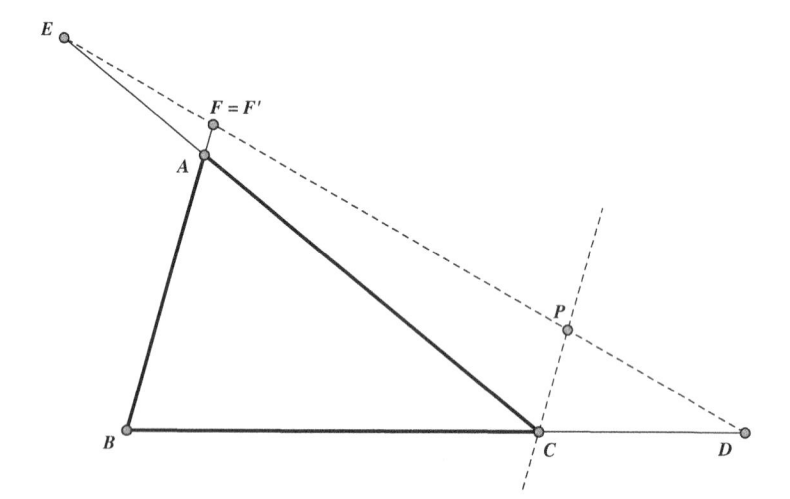

Figure 6-8

Here, we have triangle ABC with three points, D, E, and F, on the extended sides BC, CA, and AB, respectively. In this configuration, we can also show that the three points D, E, and F are collinear if, and only if, $AF \cdot BD \cdot CE = AE \cdot BF \cdot CD$. The proof in this case is identical to that previously shown. We see that Menelaus' theorem can be stated in a more general fashion as follows: Given a triangle ABC, where points D, E, and F lie on the lines BC, CA, and AB, respectively, these points are collinear if, and only if, $AF \cdot BD \cdot CE = AE \cdot BF \cdot CD$, or $\frac{AE}{CE} \cdot \frac{BF}{AF} \cdot \frac{CD}{BD} = 1$.

Ceva's Theorem

A common topic in triangle geometry concerns lines joining the vertices of a triangle with points on their opposite sides. Unfortunately, school mathematics does not consider such lines beyond altitudes, angle bisectors, and medians of the triangle. Such lines are often called *Cevians*, named after the Italian mathematician Giovanni Ceva (1647–1734). Ceva's theorem, published in 1678, is very powerful and leads to many geometric wonders, which are unfortunately not shown in the school curriculum. Ceva's theorem states that three Cevians are concurrent if, and only if, the product of the lengths of the alternate segments made by the points of contact on the sides are equal. We see this illustrated in Figure 6-9, where triangle ABC has Cevians AD, BE, and CF, meeting the sides BC, CA, and AB at points D, E, and F, respectively, and are shown to meet at a common point P. According to Ceva's theorem, the Cevians BC, CA, and AB intersect in a common point P if, and only if, $AF \cdot BD \cdot CE = AE \cdot BF \cdot CD$, or equivalently, $\frac{AE}{CE} \cdot \frac{BF}{AF} \cdot \frac{CD}{BD} = 1$.

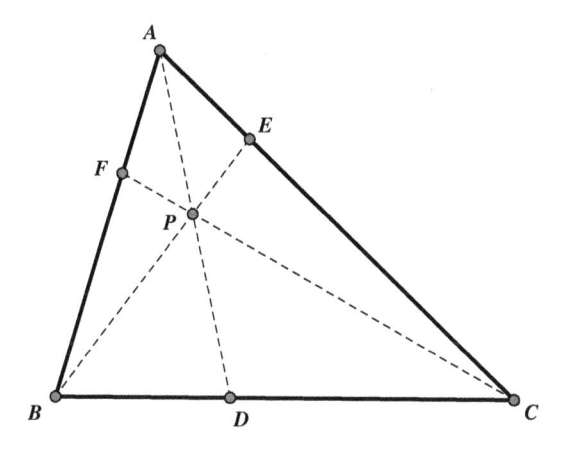

Figure 6-9

Proof: We will divide this proof into steps. First, we will assume that the three Cevians, BC, CA and AB, intersect at a common point P, and we will prove that $\frac{AE}{CE} \cdot \frac{BF}{AF} \cdot \frac{CD}{BD} = 1$. To the configuration shown in Figure 6-9, we have added a line parallel to BC through A, which intersects BE extended at point R and CF extended at point S, as shown in Figure 6-10.

The parallel lines enable us to establish several pairs of similar triangles, and the relationships between the lengths of their sides resulting from these similarities. Thus, we have

$$\triangle AER \sim \triangle CEB \text{, and therefore, } \frac{AE}{CE} = \frac{AR}{BC}, \tag{I}$$

$$\triangle BCF \sim \triangle ASF \text{, and therefore, } \frac{BF}{AF} = \frac{BC}{AS}, \tag{II}$$

$$\triangle CPD \sim \triangle SPA \text{, and therefore, } \frac{CD}{AS} = \frac{DP}{AP}, \tag{III}$$

$$\triangle BDP \sim \triangle RAP \text{, and therefore, } \frac{BD}{AR} = \frac{DP}{AP}. \tag{IV}$$

From (III) and (IV) we get: $\dfrac{CD}{AS} = \dfrac{BD}{AR}$, or $\dfrac{CD}{BD} = \dfrac{AS}{AR}$. $\tag{V}$

Now by multiplying (I), (II), and (V), we obtain our desired result:

$$\frac{AE}{CE} \cdot \frac{BF}{AF} \cdot \frac{CD}{BD} = \frac{AR}{BC} \cdot \frac{BC}{AS} \cdot \frac{AS}{AR} = 1.$$

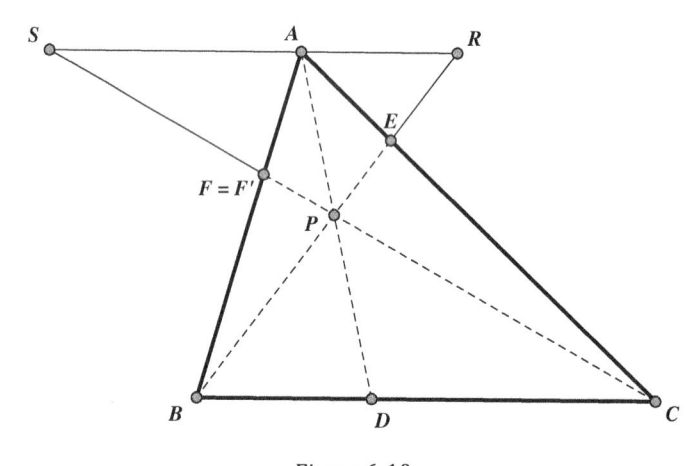

Figure 6-10

Next, we will assume that points D, E, and F are given on the triangle sides BC, CA, and AB, respectively, where

$$\frac{AE}{CE} \cdot \frac{BF}{AF} \cdot \frac{CD}{BD} = 1.$$

We wish to prove that line segments *AD*, *BE*, and *CF* are concurrent, that is, they have a common point *P*. Suppose *AD* and *BE* intersect at *P*. Draw *CP* and call its intersection with *AB* point *F'*. Since *AD*, *BE*, and *CF'* are concurrent, we can use the part of Ceva's theorem we have already proved to state the following:

$$\frac{AE}{CE} \cdot \frac{BF'}{AF'} \cdot \frac{CD}{BD} = 1.$$

Since our hypothesis stated that

$$\frac{AE}{CE} \cdot \frac{BF}{AF} \cdot \frac{CD}{BD} = 1,$$

we obtain $\frac{BF'}{AF'} = \frac{BF}{AF}$, and thus $F = F'$, thereby proving the concurrency.

Ceva's Theorem Extended

Ceva's theorem can also be applied when the Cevians intersect outside the triangle. In Figure 6-11, the Cevians *AD*, *BE*, and *CF* meet the sides

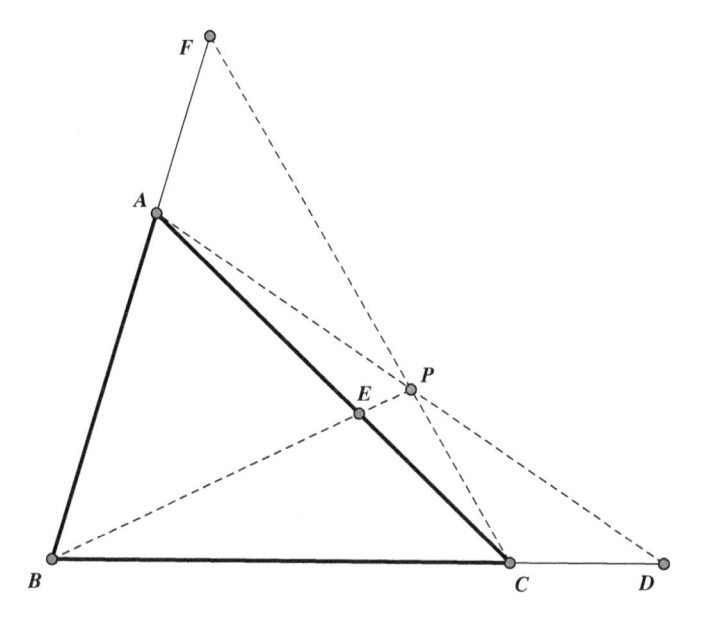

Figure 6-11

BC, *CA*, and *AB* at points *D*, *E*, and *F*, respectively. Here, the configuration is such that they intersect outside the triangle. In this case, it is also true that *AD*, *BE*, and *CF* meet in a common point *P* if, and only if,

$$\frac{AE}{CE} \cdot \frac{BF}{AF} \cdot \frac{CD}{BD} = 1.$$

Proof: The proof in this case is identical to the proof when point *P* is in the interior of *ABC*. In Figure 6-12, we have added points *R* and *S* in the same way we did in Figure 6-10, and the earlier proof is identical for this configuration.

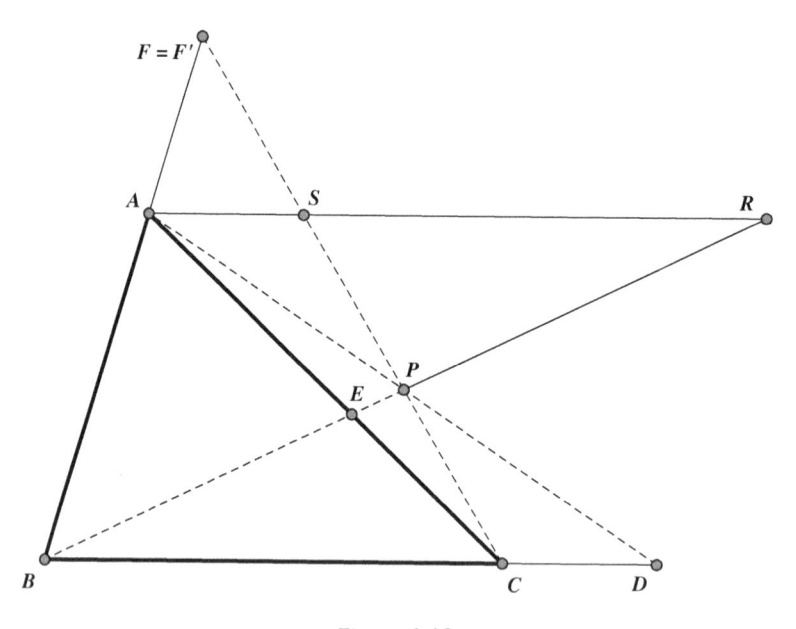

Figure 6-12

Ptolemy's Theorem

At the secondary school level, the topic of quadrilaterals is largely limited to special quadrilaterals, such as parallelograms and trapezoids. Yet, there are some amazing relationships that exist among cyclic quadrilaterals, that is, quadrilaterals that can be inscribed in a circle, where the four vertices lie on the same circle. The Alexandrian mathematician Claudius Ptolemy (c.100–c.170), who is perhaps best known for his astronomical findings, discovered an amazing geometric relationship related to the cyclic quadrilateral. The product of the lengths of the diagonals of a cyclic quadrilateral equals the sum of the products of the lengths of the pairs of opposite sides. In other words, in Figure 6-13, we have $AC \cdot BD = AB \cdot CD + BC \cdot DA$.

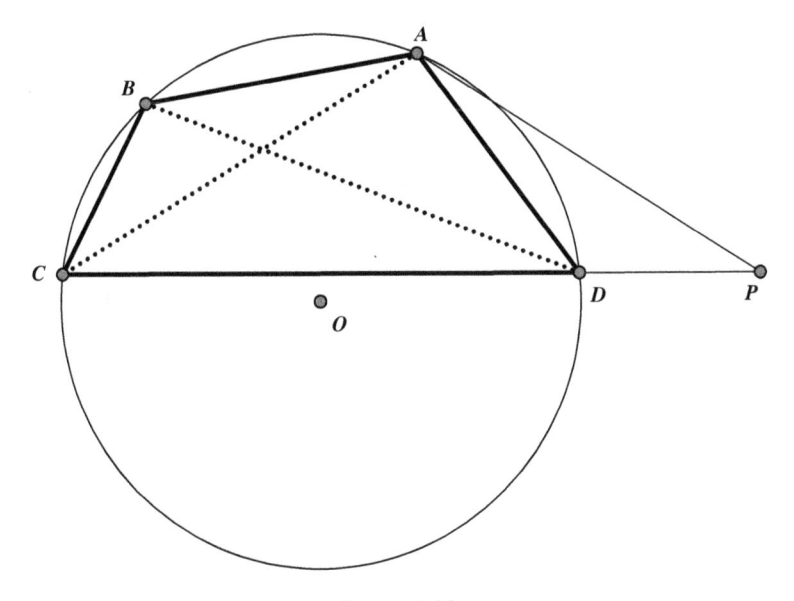

Figure 6-13

Proof: In Figure 6-13, quadrilateral $ABCD$ is inscribed in circle O. A line is drawn through A to meet CD extended at P, so that $\angle BAC = \angle DAP$. Since quadrilateral $ABCD$ is cyclic, $\angle CBA$ is supplementary to $\angle ADC$. Since $\angle PDA$ is also supplementary to $\angle ADC$, we have $\angle CBA = \angle PDA$, and because of the way point P was constructed, triangles BCA and DPA are similar. We then have $\frac{AB}{DA} = \frac{BC}{DP}$, or $DP = \frac{BC \cdot DA}{AB}$. Furthermore, we also get $\frac{AB}{DA} = \frac{AC}{AP}$. Since $\angle BAD = \angle BAC + \angle CAD = \angle DAP + \angle CAD = \angle CAP$, triangles ABD and ACP are also similar. We then have $\frac{BD}{CP} = \frac{AB}{AC}$, or $CP = \frac{AC \cdot BD}{AB}$. We know that $CP = CD + DP$, and substituting for CP and DP, we get $\frac{AC \cdot BD}{AB} = CD + \frac{BC \cdot DA}{AB}$ or $AC \cdot BD = AB \cdot CD + BC \cdot DA$, which is Ptolemy's theorem.

Desargues' Theorem

The placement of triangles in the plane is typically not considered at the high school level. However, a very useful geometric relationship in this regard, discovered by French mathematician Gérard Desargues (1591–1661), was first presented in 1648 in a book entitled *Manière universelle de M. Desargues, pour pratiquer la perspective.* It involves two triangles placed so that the three lines joining corresponding vertices are concurrent. Remarkably, when this is the case, the pairs of corresponding sides intersect in three collinear points. In Figure 6-14, we have $\triangle ABC$ and $\triangle A'B'C'$ situated in such a way that the lines joining the corresponding vertices AA', BB', and CC' are concurrent. The pairs of corresponding sides, therefore, intersect in three collinear points. In other words, point Q, in which BC and $B'C'$ intersect, point R, in which CA and $C'A'$ intersect, and point S, in which AB and $A'B'$ intersect, are collinear. As we would expect, the converse is also true: If the two triangles ABC and $A'B'C'$ are situated in such a way that points Q, R, and S lie on a common line, lines AA', BB', and CC' contain a common point P.

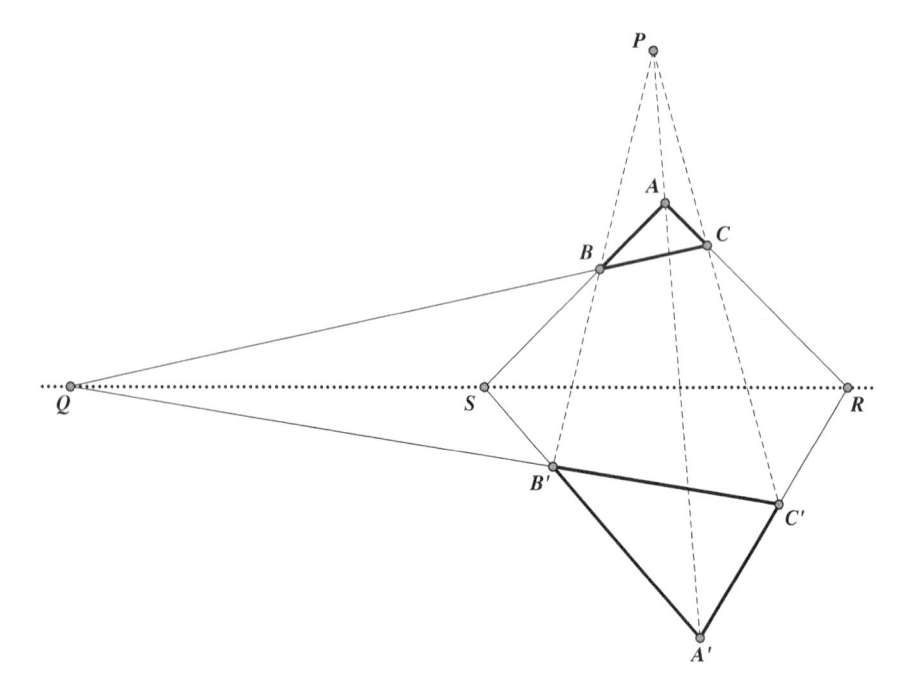

Figure 6-14

Proof: We shall prove Desargues' theorem by applying Menelaus' theorem several times. Consider line QBC as a transversal of $\triangle PB'C'$. By Menelaus' theorem, we have

$$\frac{PB}{BB'} \cdot \frac{B'Q}{QC'} \cdot \frac{C'C}{CP} = 1. \tag{I}$$

Similarly, considering SBA as a transversal of $\triangle PB'A'$, we have

$$\frac{PA}{AA'} \cdot \frac{A'S}{SB'} \cdot \frac{B'B}{BP} = 1, \tag{II}$$

and considering RCA as a transversal of $\triangle PA'C'$, we have

$$\frac{PC}{CC'} \cdot \frac{C'R}{RA'} \cdot \frac{A'A}{AP} = 1. \tag{III}$$

By multiplying (I), (II), and (III), we get

$$\left(\frac{PB}{BB'} \cdot \frac{B'Q}{QC'} \cdot \frac{C'C}{CP} \right) \left(\frac{PA}{AA'} \cdot \frac{A'S}{SB'} \cdot \frac{B'B}{BP} \right) \left(\frac{PC}{CC'} \cdot \frac{C'R}{RA'} \cdot \frac{A'A}{AP} \right) = 1,$$

or

$$\frac{B'Q}{QC'} \cdot \frac{A'S}{SB'} \cdot \frac{C'R}{RA'} = 1.$$

Thus, when Menelaus' theorem is applied to $\triangle A'B'C'$, we have points A', B', and C' collinear. It now remains for us to prove that the converse is also true. We assume that ABC and $A'B'C'$ are situated such that points Q, R, and S lie on a common line, and wish to show that lines AA', BB', and CC' contain a common point P. Surprisingly, this is an immediate consequence of what we have just proved. Consider triangles $SB'B$ and RCC'. Because of our assumptions, we know that lines BC, $B'C'$, and RS have a common point, namely, Q. From the version of Desargues' theorem that we have just proved, we know that the pairs of corresponding sides of $SB'B$ and RCC' intersect in three collinear points. In other words, the point P, in which BB' and CC' intersect, the point A, in which BS and CR intersect, and the point A', in which $B'S$ and $C'R$ intersect, are collinear. Lines AA', BB', and CC', therefore, all pass through the common point P.

Simson's Theorem

As mentioned earlier, the collinearity of points is not addressed properly at the secondary school level. Consequently, most high school geometry students are not familiar with the simple and practical relationship of collinear points known as Simson's theorem. Yet, this theorem was actually developed by William Wallace (1768–1843), and published in Thomas Leybourn's *Mathematical Repository* in 1799. Through careless misquotes, this theorem has been attributed to the famous Scottish mathematician Robert Simson (1687–1768), who interpreted Euclid's *Elements*, a foundational text for the study of geometry for centuries. Simson's theorem (to use the popular name) states that the feet of the perpendiculars drawn from any point on the circumscribed circle of a triangle to the sides of the triangle are collinear. This is shown in Figure 6-15, where point P is a point on the circumscribed circle of triangle ABC. Drawing $PX \perp BC$, $PY \perp CA$, and $PZ \perp AB$, (with X on BC, Y on CA, and Z on AB) points X, Y, and Z are collinear, regardless of where point P is positioned on the circumscribed circle of the triangle. The line XZY is referred to as the *Simson line*.

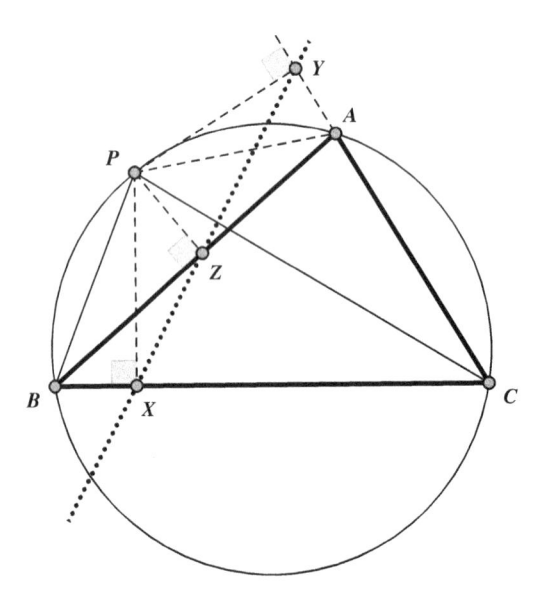

Figure 6-15

Proof: We begin by drawing *PA*, *PB*, and *PC* in Figure 6-15. Since *A*, *B*, *C*, and *P* lie on a common circle, we have $\angle ABP = \angle ACP$. Right triangles *ZPB* and *YPC* are, therefore, similar, and we have $\frac{BZ}{CY} = \frac{PZ}{PY}$. Analogously, $\angle PAB = \angle PCB$ gives us similar right triangles *XCP* and *ZAP*, so that $\frac{CX}{AZ} = \frac{PX}{PZ}$. Since $\angle CBP$ and $\angle PAC$ are opposite angles of a cyclic quadrilateral, they are supplementary. However, $\angle YAP$ is also supplementary to $\angle PAC$. This means that right triangles *XPB* and *YPA* are also similar, giving us $\frac{AY}{BX} = \frac{PY}{PX}$. Multiplying these identities, we obtain

$$\frac{BZ}{CY} \cdot \frac{CX}{AZ} \cdot \frac{AY}{BX} = \frac{PZ}{PY} \cdot \frac{PX}{PZ} \cdot \frac{PY}{PX} = 1.$$

Thus, by Menelaus' theorem we know that the points *X*, *Y*, and *Z* are collinear.

Heron's Formula

Typically, high school teaches students that the area of a triangle is determined by half the product of the lengths of its altitude and base. A more advanced formula, where the area of a triangle is equal to one-half of the product of two sides and the sine of the included angle, is also sometimes taught. Unfortunately, a most amazing method to find the area of a triangle using only the lengths of the sides is typically not presented to the secondary school population. We know that the angles of a triangle are uniquely determined if the lengths of all three sides are given. This means that all triangles with equal corresponding side lengths are congruent, and thus, they have equal areas. Heron's formula allows us to calculate the area of a triangle using only the lengths of its three sides. If triangle *ABC* has sides of length *a*, *b*, and *c*, where the semiperimeter $s = \frac{a+b+c}{2}$, Heron's formula states that

$$\text{area}\triangle ABC = \sqrt{s(s-a)(s-b)(s-c)}$$
$$= \sqrt{\frac{a+b+c}{2} \cdot \frac{-a+b+c}{2} \cdot \frac{a-b+c}{2} \cdot \frac{a+b-c}{2}}.$$

Proof: There are many ways to prove this. The following method is perhaps the most basic, as it only requires the Pythagorean theorem and some basic algebraic manipulation. As we see in Figure 6-16, we are given a triangle ABC with sides a, b, and c as shown. The vertices are named in such a way that there exists a point D on side AB, with $CD \perp AB$.

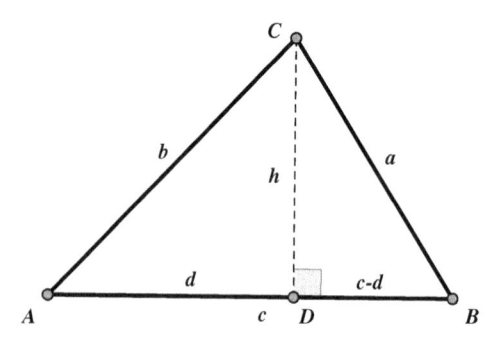

Figure 6-16

In right triangles DCA and DBC, the Pythagorean theorem gives us $b^2 = d^2 + h^2$ and $a^2 = (c-d)^2 + h^2 = d^2 - 2cd + c^2 + h^2$, respectively, and subtraction then yields $a^2 - b^2 = c^2 - 2cd$ or $d = \frac{-a^2 + b^2 + c^2}{2c}$. We can now express the altitude h by

$$h^2 = b^2 - d^2$$

$$h^2 = (b+d)(b-d)$$

$$h^2 = \left(b + \frac{-a^2 + b^2 + c^2}{2c}\right)\left(b - \frac{-a^2 + b^2 + c^2}{2c}\right)$$

$$h^2 = \frac{2bc - a^2 + b^2 + c^2}{2c} \cdot \frac{2bc + a^2 - b^2 - c^2}{2c}$$

$$h^2 = \frac{\left((b+c)^2 - a^2\right) \cdot \left(a^2 - (b-c)^2\right)}{4c^2}$$

$$h^2 = \frac{(a+b+c)(-a+b+c)(a-b+c)(a+b-c)}{4c^2}$$

$$h^2 = \frac{2s \cdot 2(s-a) \cdot 2(s-b) \cdot 2(s-c)}{4c^2}$$

$$h^2 = \frac{4s(s-a)(s-b)(s-c)}{c^2}.$$

This allows us to calculate the area of the triangle ABC as

$$\text{area}\triangle ABC = \frac{1}{2} \cdot ch = \sqrt{\frac{c^2}{4} \cdot \frac{4s(s-a)(s-b)(s-c)}{c^2}} = \sqrt{s(s-a)(s-b)(s-c)}.$$

The Gergonne Point

Just as collinearity is not given proper attention in the high school curriculum, so too is the concurrency of three or more lines often overlooked. A fascinating point of concurrency in a triangle was first established by Joseph-Diaz Gergonne (1771–1859), a French mathematician who earned a distinct place in the history of mathematics as the founder (1810) of the first purely mathematical journal, *Annales des mathématiques pures et appliqués.* The journal was published monthly until 1832 and was known as *Annales del Gergonne.* During its run, Gergonne published about 200 papers, mostly on geometry. Gergonne's *Annales* played an important role in the establishment of projective and algebraic geometry, as it gave some of the greatest minds of the times an opportunity to share their findings. We shall consider a rather simple theorem established by Gergonne, since it exhibits concurrency and is easily proved using Ceva's theorem.

Gergonne's theorem states that the lines containing a vertex of a triangle and the point of tangency of the opposite side with the inscribed circle are concurrent. This point of concurrency is known as the *Gergonne point* of the triangle.

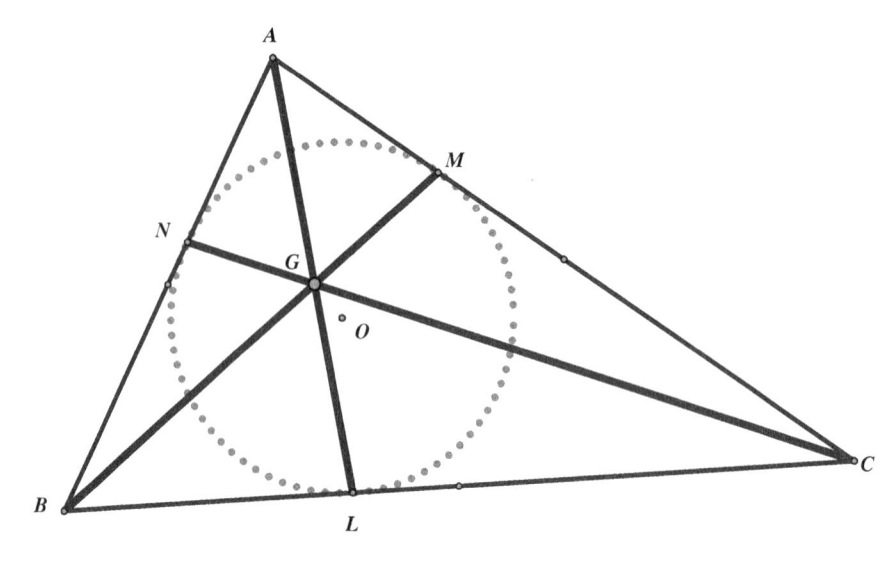

Figure 6-17

Proof: In Figure 6-17, circle O is tangent to sides AB, AC, and BC in points N, M, and L, respectively. It follows that $AN = AM$, $BL = BN$, and $CM = CL$. These equalities may be written as $\frac{AN}{AM} = 1$, $\frac{BL}{BN} = 1$, and $\frac{CM}{CL} = 1$. By multiplying these three fractions we get $\frac{AN}{AM} \cdot \frac{BL}{BN} \cdot \frac{CM}{CL} = 1$. Therefore, $\frac{AN}{BN} \cdot \frac{BL}{CL} \cdot \frac{CM}{AM} = 1$, which, by applying Ceva's theorem, implies that AL, BM, and CN are concurrent at the point G, which is the Gergonne point of $\triangle ABC$.

Pascal's Theorem

Most inscribed polygons encountered at the secondary school level are regular polygons. Yet, inscribed polygons which are not regular also produce some amazing relationships. Blaise Pascal (1623–1662), a contemporary of Desargues and Descartes, is regarded today as one of the true geniuses in the history of mathematics. Although his eccentricities prevented him from achieving the fame he deserves for his contributions to other branches of mathematics, our concern here is one of his contributions to geometry.

In 1640, at the age of sixteen, Pascal published a one-page paper entitled *Essay pour les coniques*. It contained a theorem that Pascal referred to as *mysterium hexagrammicum*. The work highly impressed Descartes, who couldn't believe it was the work of a boy. It states that the intersections of the opposite sides of a hexagon inscribed in a conic section are collinear. For our purposes, we shall consider only the case where the conic section is a circle, and the hexagon has *no* pair of opposite sides parallel. Pascal's theorem states that if a hexagon, with no pair of opposite sides parallel, is inscribed in a circle, then the intersections of the opposite sides are collinear.

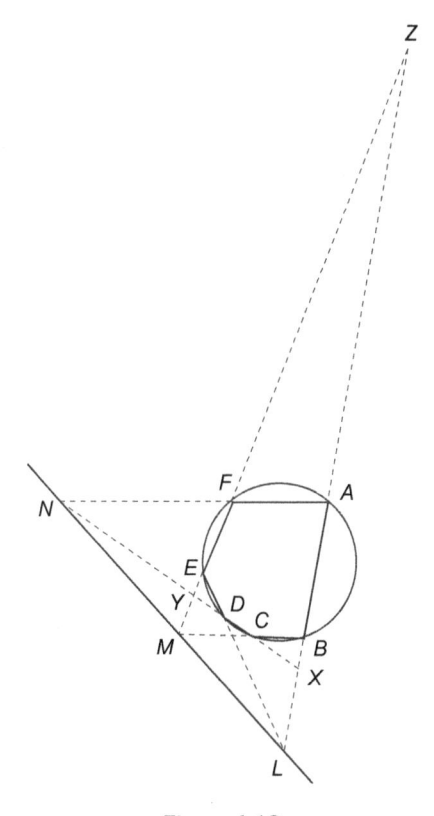

Figure 6-18

Proof: In Figure 6-18, hexagon *ABCDEF* is inscribed in a circle. The pairs of opposite sides *AB* and *DE* meet at *L*, *CB* and *EF* meet at *M*, and *CD* and *AF* meet at *N*. Also, *AB* meets *CN* at *X*, *EF* meets *CN* at *Y*, and *EF* meets *AB* at *Z*.

Consider BC to be a transversal of $\triangle XYZ$.

Then, $\dfrac{ZB}{BX} \cdot \dfrac{XC}{CY} \cdot \dfrac{YM}{MZ} = -1$ (Menelaus' theorem). (I)

Now taking AF to be a transversal of $\triangle XYZ$:

$$\dfrac{ZA}{AX} \cdot \dfrac{YF}{FZ} \cdot \dfrac{XN}{NY} = -1, \text{ by Menelaus' theorem.} \qquad \text{(II)}$$

Also, since DE is a transversal of $\triangle XYZ$:

$$\dfrac{XD}{DY} \cdot \dfrac{YE}{EZ} \cdot \dfrac{ZL}{LX} = -1, \text{ by Menelaus' theorem.} \qquad \text{(III)}$$

By multiplying (I), (II), and (III), we get:

$$\dfrac{YM}{MZ} \cdot \dfrac{XN}{NY} \cdot \dfrac{ZL}{LX} \cdot \dfrac{(ZB)(ZA)}{(EZ)(FZ)} \cdot \dfrac{(XD)(XC)}{(AX)(BX)} \cdot \dfrac{(YE)(YF)}{(DY)(CY)} = -1. \qquad \text{(IV)}$$

When two secant segments are drawn to a circle from an external point, the product of the lengths of one secant and its external segment equals the product of the lengths of the other secant and its external segment. Therefore:

$$\dfrac{(ZB)(ZA)}{(EZ)(FZ)} = 1, \qquad \text{(V)}$$

$$\dfrac{(XD)(XC)}{(AX)(BX)} = 1, \qquad \text{(VI)}$$

and $\dfrac{(YE)(YF)}{(DY)(CY)} = 1.$ (VII)

By substituting (V), (VI), and (VII) into (IV), we get:

$$\dfrac{YM}{MZ} \cdot \dfrac{XN}{NY} \cdot \dfrac{ZL}{LX} = -1.$$

Thus, by Menelaus' theorem, points M, N, and L must be collinear. This theorem can be extended in the following manner.

Variation on Pascal's Theorem

A variation of Pascal's theorem states that if a hexagon has its vertices on a circle in any order, then the intersections (if they exist) of the opposite sides are collinear. As an example of this variation, consider the above proof using the diagram in Figure 6-19. Only one minor adjustment needs to be made, and that is the reason for equation (V) through (VII). Remember, the same pairs of "opposite sides" are used here as were used earlier.

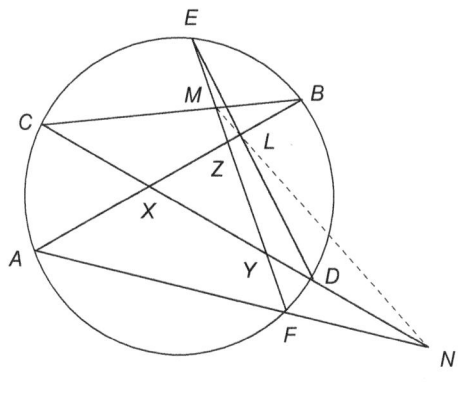

Figure 6-19

Pascal's theorem has many applications. We shall only consider a couple.

Pascal Application 1: In Figure 6-20, point P is any point in the interior of $\triangle ABC$. Points M and N are the feet of the perpendiculars from P to AB and AC, respectively. Furthermore, $AK \perp CP$ at K and $AL \perp BP$ at L. We then have that KM, LN, and BC are concurrent.

Proof: We can prove that points A, K, M, P, N, and L all lie on the circle with diameter AP by realizing that right angles AKP and AMP are

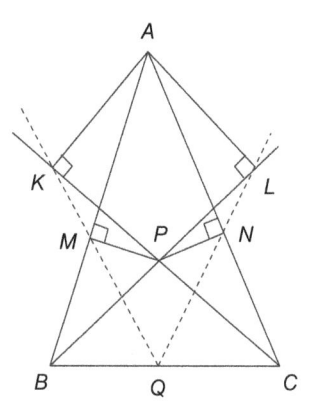

Figure 6-20

inscribed in the same semicircle, as is the case for right angles *ALP* and *ANP*. Now, using the variation to Pascal's theorem previously considered, we notice that for inscribed hexagon *AKMPNL*, the pairs of sides intersect as follows:

$$AM \cap LP = B,$$
$$AN \cap KP = C,$$
$$KM \cap LN = Q.$$

By Pascal's theorem, *B*, *C*, and *Q* are collinear, which is to say that *KM*, *LN*, and *BC* are concurrent.

Pascal Application 2: Select any point *P* not on △*ABC* and draw a line *ℓ* that contains *P* and intersects sides *BC*, *AB* and *AC* at points *X*, *Y*, and *Z*, respectively, as shown in Figure 6-21. Let *AP*, *BP*, and *CP* intersect the circumcircle of △*ABC* at points *R*, *S*, and *T*, respectively. We then have lines *RX*, *SZ*, and *TY* concurrent.

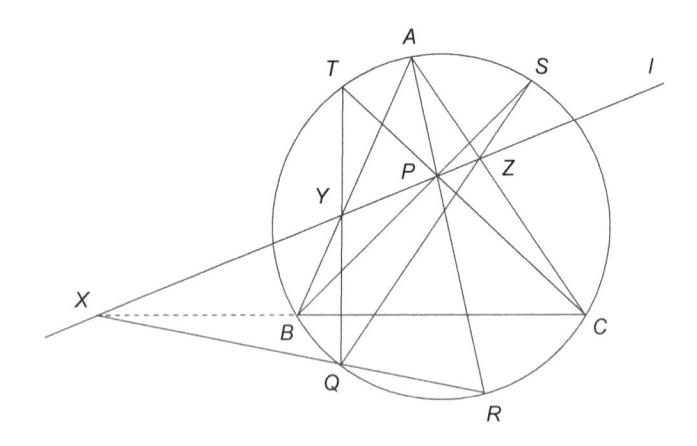

Figure 6-21

Proof: Let *RX* intersect the circumcircle at *Q*. Consider hexagon *ARQTCB*, and apply Pascal's theorem to it. We notice that since *AR* ∩ *TC* at *P*, and *RQ* ∩ *CB* at *X*, then *TQ* ∩ *AB* at a point on *ℓ*, which must be *Y* (since *AB* ∩ *ℓ* at *Y*). Now consider hexagon *ARQSBC*. Similarly, since *AR* ∩ *SB* at *P*, and *RQ* ∩ *CB* at *X*, then *SQ* ∩ *AC* at a point on *ℓ*, which must be *Z*. Thus, *RX*, *SZ*, and *TY* are concurrent.

Brianchon's Theorem

Instead of considering a polygon inscribed in a circle as we did previously, we will consider a polygon circumscribed about a circle. This is another topic that's not traditionally part of the secondary school curriculum, though it's easily comprehended at that level. In 1806, at the age of twenty-one, a student at the École Polytechnique, Charles Julien Brianchon (1785–1864) published an article in the *Journal de L'École Polytechnique* that became one of the fundamental contributions to the study of conic sections in projective geometry. His findings led to a restatement of Pascal's theorem and its extension. Brianchon's theorem states that in any hexagon circumscribed about a conic section, the three diagonals intersect each other in the same point. Brianchon's theorem bears a curious resemblance to Pascal's theorem, and they are, in fact, duals of one another. The table below compares the two theorems; notice that the two statements are alike except for the underlined words, which are duals of one another. As with Pascal's theorem, we shall consider only the conic section that is a circle.

Pascal's theorem	Brianchon's theorem
The <u>points of intersection</u> of the opposite <u>sides</u> of a hexagon <u>inscribed in</u> a conic section are <u>collinear</u>.	The <u>lines joining</u> the opposite <u>vertices</u> of a hexagon <u>circumscribed about</u> a conic section are <u>concurrent</u>.

Theorem: If a hexagon is circumscribed about a circle, as shown in Figure 6-22, the lines containing opposite vertices are concurrent.

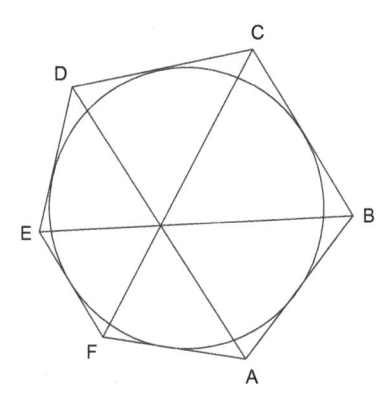

Figure 6-22

Proof: As seen in Figure 6-23, the sides of hexagon *ABCDEF* are tangent to a circle at points *T*, *N*, *L*, *S*, *M*, and *K*. Points *K′*, *L′*, *N′*, *M′*, *S′*, and *T′* are chosen on *FA*, *DC*, *BC*, *FE*, *DE*, and *BA*, respectively, so that

$$KK' = LL' = NN' = MM' = SS' = TT'.$$

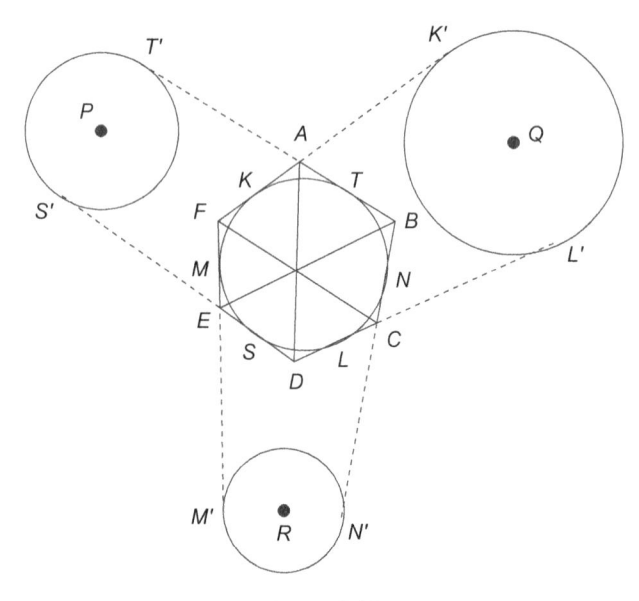

Figure 6-23

Construct circle *P* tangent to *BA* and *DE* at points *T′* and *S′*, respectively (the existence of this circle is easily justified). Similarly, construct circle *Q* tangent to *FA* and *DC* at points *K′* and *L′*, respectively. Then construct circle *R* tangent to *FE* and *BC* at points *M′* and *N′*, respectively. Since two tangent segments to a circle from an external point have the same length, *FM* = *FK*. We already know that *MM′* = *KK′*. Therefore, by addition *FM′* = *FK′*. Similarly, *CL* = *CN* and *LL′* = *NN′*. By subtraction *CL′* = *CN′*. We now notice that points *F* and *C* are endpoints of a pair of congruent tangent segments to circles *R* and *Q*. Thus, these points determine the radical axis, *CF*, of circles *R* and *Q*. Using the same technique, we can easily show that *AD* is the radical axis of circles P and Q, and that BE is the radical axis of circles P and R. We know that the radical axes of three circles with non-collinear centers (taken in pairs) are concurrent. Therefore, CF, AD, and BE are concurrent. We should note that the only way in which these circles would have had collinear centers is if the diagonals were to have coincided, which is impossible.

Theorem: Pentagon $ABCDE$ is circumscribed about a circle, with points of tangency at F, M, N, R, and S. If diagonals AD and BE intersect at P, we have that C, P, and F are collinear, as shown in Figure 6-24.

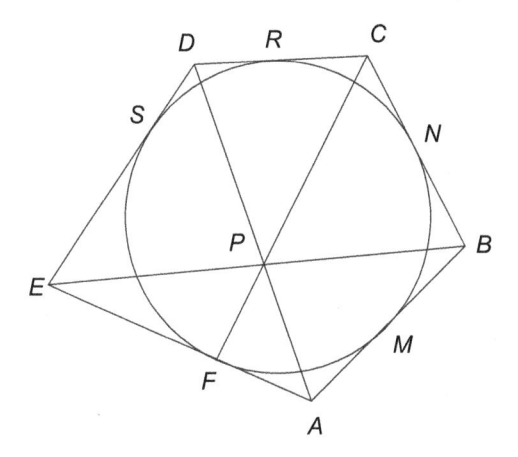

Figure 6-24

Proof: Consider the hexagon circumscribed about a circle, shown in Figure 6-22, having its sides AF and EF merge into one line segment. Thus, AFE is now a side of a circumscribed pentagon with F as one point of tangency, as seen in Figure 6-24. We can view the pentagon in Figure 6-24 as a degenerate hexagon, and then simply apply Brianchon's theorem to this degenerate hexagon to obtain our desired conclusion. That is, AD, BE and CF are concurrent at P, or C, P, and F are collinear.

Pappus' Theorem

Another geometric relationship that could easily be included in the secondary school curriculum is the following. Suppose we consider the vertices of a hexagon $AB'CA'BC'$, as seen in Figure 6-25, being located alternately on two lines, as illustrated in Figure 6-26. Suppose we now draw the lines that were the opposite sides of the hexagon to locate their point of intersection. We find that the three points of intersection of these pairs of "opposite sides" are collinear. This conclusion was first published by Pappus of Alexandria in his *Mathematical Collection* in 300 CE.

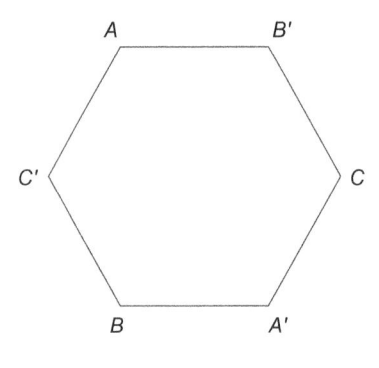

Figure 6-25

Theorem: Points A, B, and C are on one line and points A', B', and C' are on another line (in any order). If AB' and $A'B$ meet at C'', while AC' and $A'C$ meet at B'', and BC' and $B'C$ meet at A'', then points A'', B'', and C'' are collinear.

Proof: In Figure 6-26, $B'C$ meets $A'B$ at Y, AC' meets $A'B$ at X, and $B'C$ meets AC' at Z. Consider $C''AB'$ as a transversal of $\triangle XYZ$:

$$\frac{ZB'}{YB'} \cdot \frac{XA}{ZA} \cdot \frac{YC''}{XC''} = -1 \text{ (Menelaus' theorem).} \tag{I}$$

Now taking $A'B''C$ as a transversal of $\triangle XYZ$

$$\frac{YA'}{XA'} \cdot \frac{XB''}{ZB''} \cdot \frac{ZC}{YC} = -1 \text{ (Menelaus' theorem).} \tag{II}$$

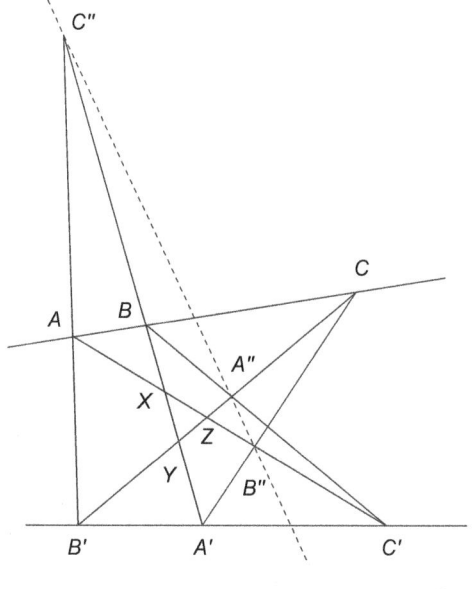

Figure 6-26

$BA''C'$ is also a transversal of $\triangle XYZ$, so that

$$\frac{YB}{XB} \cdot \frac{ZA''}{YA''} \cdot \frac{XC'}{ZC'} = -1 \text{ (Menelaus' theorem)}. \qquad \text{(III)}$$

Multiplying (I), (II), and (III) gives us equation (IV):

$$\frac{YC''}{XC''} \cdot \frac{XB''}{ZB''} \cdot \frac{ZA''}{YA''} \cdot \frac{ZB'}{YB'} \cdot \frac{YA'}{XA'} \cdot \frac{XC'}{ZC'} \cdot \frac{XA}{ZA} \cdot \frac{ZC}{YC} \cdot \frac{YB}{XB} = -1. \qquad \text{(IV)}$$

Since points A, B, and C are collinear, and A', B', and C' are collinear, we obtain the following two relationships by Menelaus' theorem when we consider each line as a transversal of $\triangle XYZ$:

$$\frac{ZB'}{YB'} \cdot \frac{YA'}{XA'} \cdot \frac{XC'}{ZC'} = -1, \qquad \text{(V)}$$

$$\frac{XA}{ZA} \cdot \frac{ZC}{YC} \cdot \frac{YB}{XB} = -1, \qquad \text{(VI)}$$

Substituting (V) and (VI) into (IV), we get:

$$\frac{YC''}{XC''} \cdot \frac{XB''}{ZB''} \cdot \frac{ZA''}{YA''} = -1.$$

Thus, points A'', B'', and C'' are collinear, by Menelaus' theorem.

Equiangular Point

In the secondary school curriculum, the basic points presented in a triangle are the centroid, the incenter, and the circumcenter. One point that is not presented is the point at which congruent angles are formed by drawing rays from this point to the triangle's vertices, as shown in Figure 6-27. Yet this point harbors many interesting relationships beyond the equality of the angles.

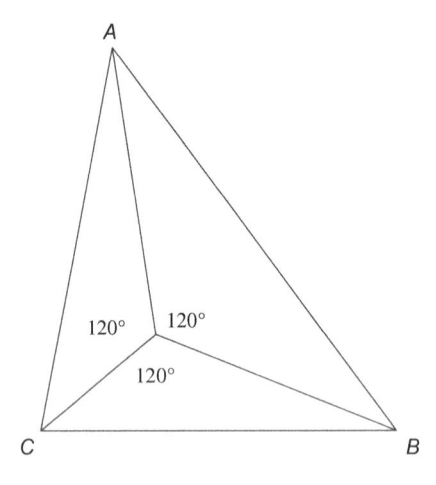

Figure 6-27

Let us now locate this point in Figure 6-27. First, we must find a point that has another interesting property. Begin by constructing an equilateral triangle externally on each side of triangle ABC. Draw segments joining each vertex of the given triangle with the remote vertex of the equilateral triangle on the opposite side, as shown in

Figure 6-28. The theorem below presents an astonishing property of these three line segments. After we prove this property, we shall return to the equiangular point mentioned above.

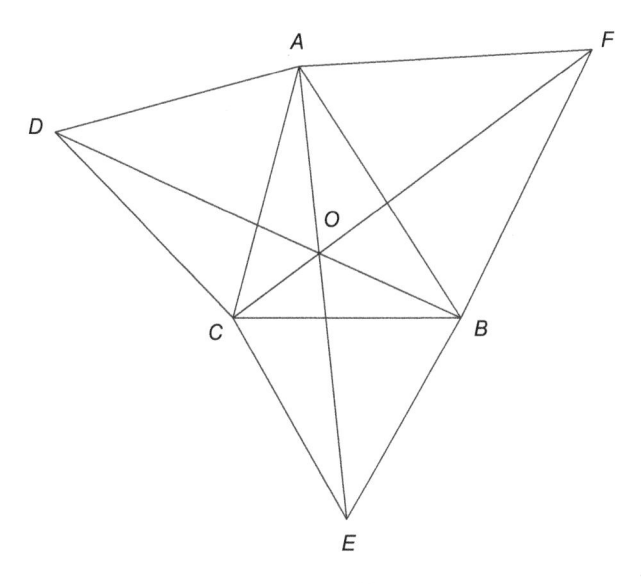

Figure 6-28

Theorem: The segments joining each vertex of a given triangle with the remote vertex of the equilateral triangle (drawn externally on the opposite side of the given triangle) are congruent.

Proof: Since $\angle DCA = \angle ECB = 60°$, $\angle DCB = \angle ACE$ (by addition). Also, since we have equilateral triangles, $DC = AC$ and $CB = CE$. Therefore, $\triangle DCB \cong \triangle ACE$ (side-angle-side) and $DB \cong AE$. In a similar manner, we may prove that $\triangle EBA \cong \triangle CBF$. This enables us to conclude that $AE = CF$. Thus, $DB = AE = CF$. From the diagram in Figure 6-28, it appears that DB, AE, and CF are concurrent. This observation generates our next theorem.

Theorem: The segments joining each vertex of a given triangle with the remote vertex of the equilateral triangle drawn externally on the opposite side of the given triangle are concurrent. This point is called the *Fermat point*[5] of the triangle.

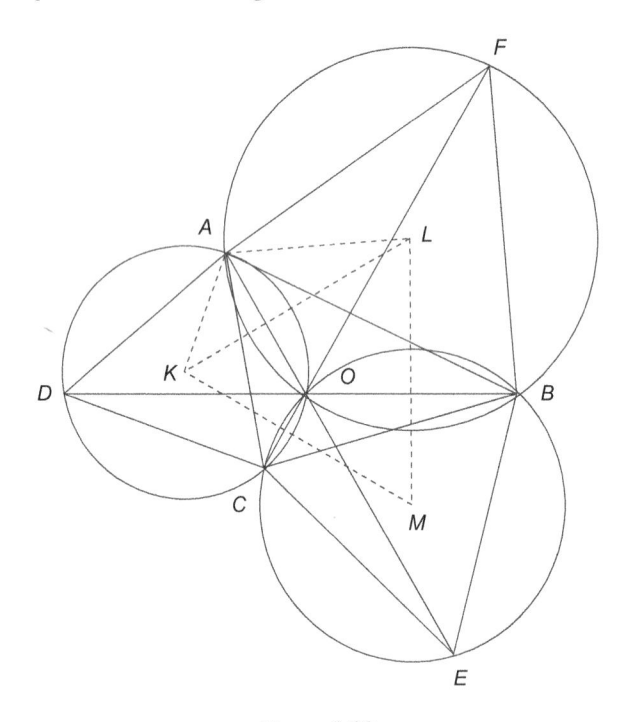

Figure 6-29

Proof: Consider the circumcircles of the three equilateral triangles $\triangle ACD$, $\triangle ABF$, and $\triangle BCE$. Let K, L, and M be the centers of these circles, as shown in Figure 6-29. Circles K and L meet at points O and A. Since $\overset{\frown}{ADC} = 240°$, and we know that $\angle AOC = \frac{1}{2}(\overset{\frown}{ADC})$, then $\angle AOC = 120°$. Similarly, $\angle AOB = \frac{1}{2}(\overset{\frown}{AFB}) = 120°$. Therefore, $m\angle COB = 120°$, since a complete revolution is $360°$. Since $\overset{\frown}{CEB} = 240°$, $\angle COB$ is an inscribed angle and point O must lie on circle M. The three circles are thus concurrent, intersecting at point O. Then join point O with points A, B, C,

---

[5] So named after the French mathematician Pierre de Fermat (1601–1665).

D, E, and F: $m\angle DOA = m\angle AOF = m\angle FOB = 60°$. Therefore, DOB is a straight line. Similarly, COF and AOE. Thus, it has been proved that DB, AE, and CF are concurrent, intersecting at point O, which is also the point of intersection of circles K, L, and M.

Can you now determine the point in $\triangle ABC$ at which the three sides subtend (i.e., determine by being opposite) congruent angles? The point O is called the *equiangular point* of $\triangle ABC$ since $m\angle AOB = m\angle AOC = m\angle BOC = 120°$.

Before continuing with the equiangular point, let us take advantage of another interesting property. Sources indicate that the following theorem was developed by Napoleon Bonaparte, who took pride in his mathematical talents. Thus, the resulting equilateral triangle is often called the *Napoleon triangle*.

Theorem: The circumcenters of the three equilateral triangles drawn externally on the sides of a given triangle determine an equilateral triangle.

Proof: Consider $\triangle DAC$ in Figure 6-29. Since K is the centroid (point of intersection of the medians) of $\triangle DAC$, AK is two-thirds of the length of the altitude (or median). Using the relationships of a 30–60–90 triangle, we find that $\frac{DB}{KM} = \frac{\sqrt{3}}{1}$, and $\frac{AE}{ML} = \frac{\sqrt{3}}{1}$. Similarly, in equilateral $\triangle AFB$, $\frac{AF}{AL} = \frac{\sqrt{3}}{1}$. Therefore, $\frac{AC}{AK} = \frac{AF}{AL}$, and $\angle KAC = \angle LAF = 30°$, $\angle CAL = \angle CAL$ (reflexive) and $\angle KAL = \angle CAF$ (addition). Therefore, $\triangle KAL \sim \triangle CAF$. Thus, $\frac{CF}{KL} = \frac{CA}{AK} = \frac{\sqrt{3}}{1}$. Similarly, we may prove $\frac{DB}{KM} = \frac{\sqrt{3}}{1}$, and $\frac{AE}{ML} = \frac{\sqrt{3}}{1}$. Therefore, $\frac{DB}{KM} = \frac{AE}{ML} = \frac{CF}{KL}$. But since $DB = AE = CF$, as proved earlier, we obtain $KM = ML = KL$. Therefore, $\triangle KML$ is equilateral.

A Property of Equilateral Triangles

Choose any convenient point in the interior region of an equilateral triangle. Now measure the distances from this point to the three sides and record the sum of these distances. (This can be done on paper or on software such as Geometer's Sketchpad or GeoGebra.) Repeat this procedure for any other point in the interior region of this triangle. How do the two sums compare? Now measure the length of the altitude of the triangle. How do the two sums compare to the length of the altitude of the equilateral triangle? The answers to these questions suggest the following theorem.

Theorem: The sum of the distances from any point in the interior of an equilateral triangle to the sides of the triangle is constant and equal to the length of the altitude of the triangle.

Proof: In equilateral $\triangle ABC$, $PR \perp AC$, $PQ \perp BC$, $PS \perp AB$, and $AD \perp BC$. Draw PA, PB and PC, as shown in Figure 6-30.

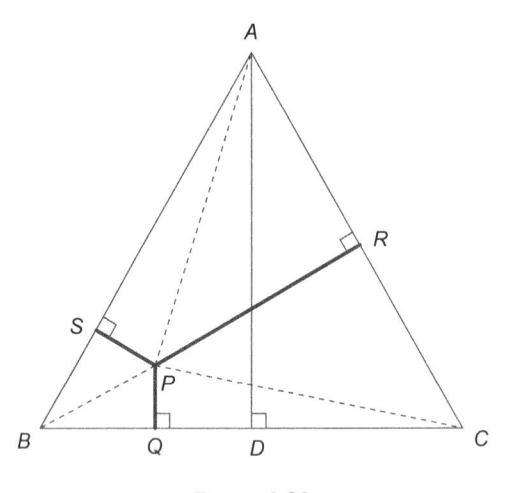

Figure 6-30

The area of $\triangle ABC$ = area of $\triangle APB$ + area of $\triangle BPC$ + area of $\triangle CPA$ = $\frac{1}{2}(AB)(PS) + \frac{1}{2}(BC)(PQ) + \frac{1}{2}(AC)(PR)$. Since $AB = BC = AC$, the area of $\triangle ABC = \frac{1}{2}(BC)[PS + PQ + PR]$. However, the area of $\triangle ABC = \frac{1}{2}(BC)(AD)$. Therefore, $PS + PQ + PR = AD$, a constant for the given triangle.

A Minimum Distance Point

The point of intersection of the diagonals of a quadrilateral is the minimum distance point of a quadrilateral, that is, the point from which the sum of the distances to each of the four vertices is minimum. To prove that among the interior points of a quadrilateral the diagonal-intersection point has the smallest sum of distances to the vertices, we simply choose any other interior point and compare its sum of distances to the vertices to that of the diagonal-intersection point. Consider quadrilateral *ABCD* in Figure 6-31 with diagonals *AC* and *BD* intersecting at *Q*.

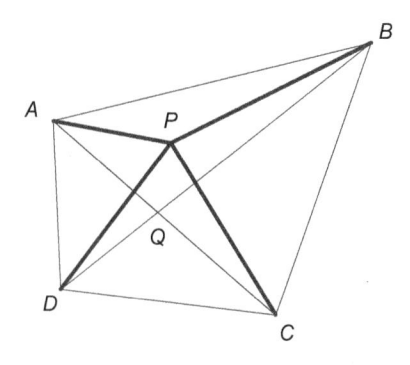

Figure 6-31

Select any point *P* (not at *Q*) in the interior of quadrilateral *ABCD*. In triangle *APC*, we have $PA + PC > QA + QC$ (since the sum of the lengths of two sides of a triangle is greater than the length of the third). Similarly, for triangle *BPD*, we have $PB + PD > QB + QD$. By addition, $PA + PB + PC + PD > QA + QB + QC + QD$, which shows that the sum of the distances from the intersection of the diagonals of a quadrilateral to the vertices is less than the sum of the distances from *any other* interior point of the quadrilateral to the vertices. This gives us the following theorem.

Theorem: The minimum distance point of a quadrilateral is the point of intersection of the diagonals.

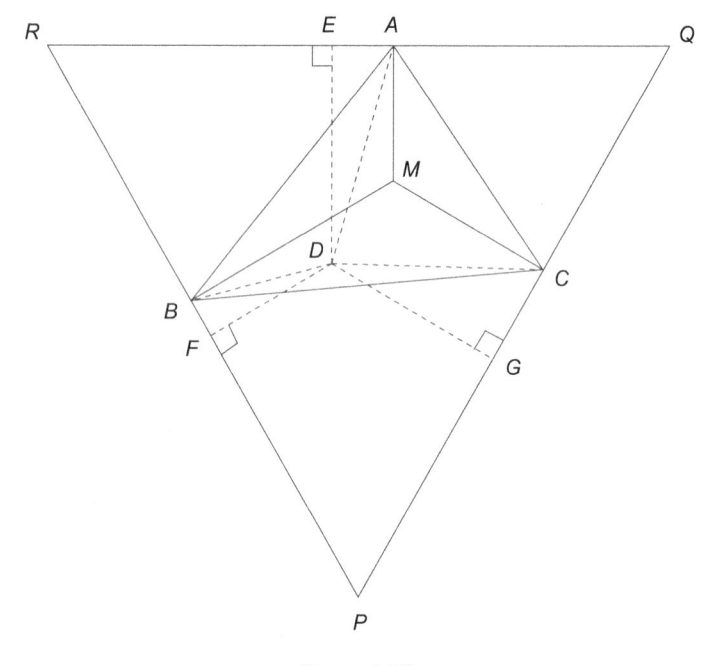

Figure 6-32

Proof: We now consider $\triangle ABC$, with no angle measuring greater than 120°, and seek the minimum distance point. Let M be the point in the interior of $\triangle ABC$, where

$$\angle AMB = \angle BMC = \angle AMC = 120°,$$

as shown in Figure 6-32. Draw lines through A, B, and C perpendicular to AM, BM, and CM, respectively. These lines meet to form equilateral $\triangle PQR$. To prove $\triangle PQR$ is equilateral, notice that each angle has measure 60°. This can be shown by considering, for example, quadrilateral $AMBR$. Since $\angle RAM = \angle RBM = 90°$, and $\angle AMB = 120°$, it follows that $\angle ARB = 60°$. Let D be *any other* point in the interior of $\triangle ABC$. We must show that the sum of the distances from M to the vertices is less than the sum of the distances from D to the vertices.

From the earlier theorem, we know that

$$MA + MB + MC = DE + DF + DG,$$

where *DE*, *DF*, and *DG* are the perpendiculars to *REQ*, *RBP*, and *QGP*, respectively.

But, *DE + DF + DG < DA + DB + DC*. Let's recall that the shortest distance from an external point to a line is the length of the perpendicular segment from the point to the line. By substitution: *MA + MB + MC < DA + DB + DC*.

We chose to restrict our discussion to triangles with angles of measure less than 120°. If you try to construct the point *M* in a triangle with one angle of measure of 150°, the reason for our restriction will become obvious. This provides the following theorem.

Theorem: The minimum distance point of a triangle (with no angle of measure greater than 120°) is the equiangular point, that is, the point at which the sides of the triangle subtend congruent angles.

Stewart's Theorem

As we mentioned earlier, the high school geometry course addresses three Cevians of a triangle: the altitude, the angle bisector, and the median. However, the general Cevian is not explored. Here we will determine how to find the length of "any" Cevian, which is a segment that has one endpoint on a vertex of a given triangle and the other endpoint on the opposite side. That is, if for $\triangle ABC$ in Figure 6-33 we know the lengths of *AC*, *BC*, *AD*, and *BD*, our challenge is to find the length of *CD*.

This problem was first solved by the famous Scottish geometer Robert Simson, who presented it in lectures. Simson allowed his notes to be used by his prize student Mathew Stewart (1717–1785) in his famous publication, *General Theorems of Considerable Use in the Higher Parts of Mathematics* (Edinburgh, 1746). Simson's generosity was motivated by his desire to see Stewart obtain the chair of mathematics at the University of Edinburgh, which he did. We shall still refer to the theorem by the author, Stewart, of the book in which it first appeared. Actually, Simson deserves particular note for his definitive book *The Elements of Euclid* (Glasgow, 1756), which was in print for over 150 years. This book is the basis for all subsequent study of Euclid's *Elements,* including the high school geometry courses taught in the United States today.

Theorem: Using the letter designations in Figure 6-33, Stewart's theorem states the following relationship: $a^2n + b^2m = c(d^2 + mn)$.

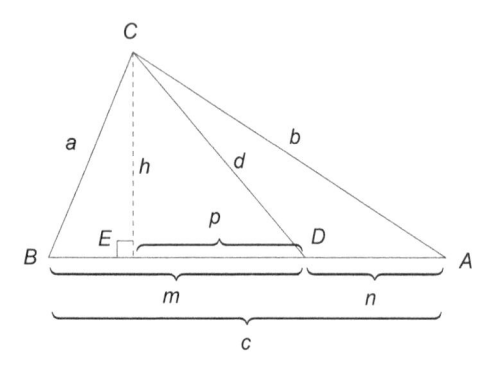

Figure 6-33

Proof: In $\triangle ABC$, let $BC = a$, $AC = b$, $AB = c$, and $CD = d$. Point D divides AB into two segments: $BD = m$, and $DA = n$. Draw altitude $CE = h$ and let $ED = p$. In order to proceed with the proof of Stewart's Theorem, we first derive two necessary formulas. The first one is applicable to $\triangle CBD$. We apply the Pythagorean theorem to $\triangle CEB$ to obtain $(CB)^2 = (CE)^2 + (BE)^2$.

$$\text{Since } BE = m - p, \quad a^2 = h^2 + (m - p)^2. \tag{I}$$

We apply the Pythagorean theorem to $\triangle CED$ to get $(CD)^2 = (CE)^2 + (ED)^2$, or $h^2 = d^2 - p^2$.
Replacing h^2 in equation (I), we obtain

$$a^2 = d^2 - p^2 + (m - p)^2,$$

$$a^2 = d^2 - p^2 + m^2 - 2mp + p^2. \tag{II}$$

$$\text{Thus, } a^2 = d^2 + m^2 - 2mp.$$

A similar argument is applicable to ΔCDA. By applying the Pythagorean theorem to ΔCEA, we find that $(CA)^2 = (CE)^2 + (EA)^2$.

Since $EA = (n + p), \quad b^2 = h^2 + (n + p)^2.$ (III)

However, $h^2 = d^2 - p^2$, so we substitute for h^2 in (III) as follows:

$$b^2 = d^2 - p^2 + (n + p)^2,$$
$$b^2 = d^2 - p^2 + n^2 + 2np + p^2.$$ (IV)
$$\text{Thus, } b^2 = d^2 + n^2 + 2np.$$

Equations (II) and (IV) give us the formulas we need. Now multiply equation (II) by n to get

$$a^2n = d^2n + m^2n - 2mnp,$$ (V)

and multiply equation (IV) by m to get

$$b^2m = d^2m + n^2m + 2mnp.$$ (VI)

Adding (V) and (VI), we have

$$a^2 + b^2m = d^2n + d^2m + m^2n + n^2m + 2mnp - 2mnp.$$

Therefore, $a^2n + b^2m = d^2(n + m) + mn(m + n)$. Since $m + n = c$, we have $a^2n + b^2m = d^2c + mnc$, or $a^2n + b^2m = c(d^2 + mn)$, which is the relationship we set out to establish.

Miquel's Theorem

This surprising relationship of circles with a triangle could easily fit into the high school curriculum. Consider any convenient triangle and select a point on each side. Then construct three circles, each containing two of these points and the vertex determined by the two sides on which these points lie. Although you can do this on paper with the aid of a pair of compasses, it is particularly nice to do this with Geometer's Sketchpad or GeoGebra. Your observation should lead you to a theorem published in 1838 by the French mathematician Auguste Miquel (1816–1851).

Theorem: If a point is selected on each side of a triangle, then the three circles determined by each vertex and the points on the adjacent sides pass through a common point.

This theorem may be viewed in two ways. The expected form is shown in Figure 6-34. However, when two of the selected points are on the extensions of the sides, the theorem still holds. This is shown in Figure 6-35.

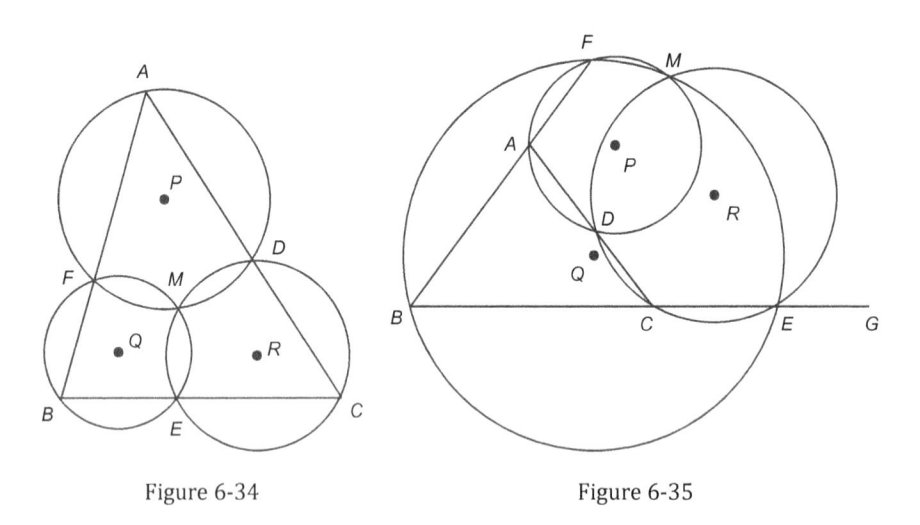

Figure 6-34 Figure 6-35

Proof:

Case I: Consider the situation when M is inside $\triangle ABC$, as shown in Figure 6-36. Points D, E, and F are any points on sides AC, BC, and AB, respectively, of $\triangle ABC$. Let circles Q and R, determined by points F, B, E and D, C, E, respectively, meet at M. Draw FM, ME, and MD.

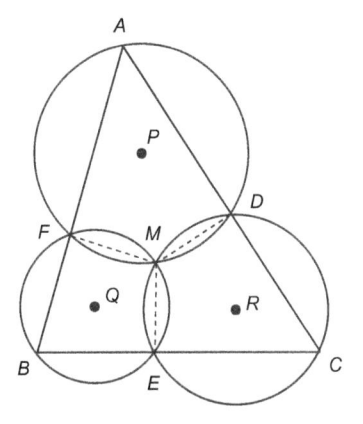

Figure 6-36

In cyclic quadrilateral *BFME*, $m\angle FME = 180° - m\angle B$. Similarly, in cyclic quadrilateral *CDME*, $\angle DME = 180° - \angle C$. By addition, $\angle FME + \angle DME = 360° - (\angle B + \angle C)$.

Therefore, $\angle FMD = \angle B + \angle C$. However, in $\triangle ABC$, $\angle B + \angle C = 180° - \angle A$.

Therefore, $\angle FMD = 180° - \angle A$ and quadrilateral *AFMD* is cyclic. Thus, point *M* lies on all three circles.

Case II: Figure 6-37 illustrates the problem when *M* is outside $\triangle ABC$.

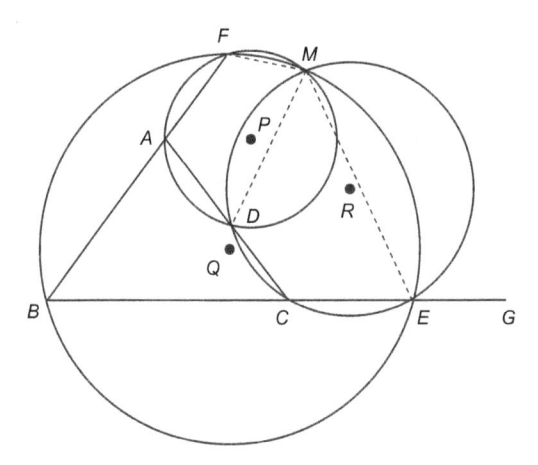

Figure 6-37

Again, let circles Q and R meet at M. Since quadrilateral $BFME$ is cyclic, $\angle FME = 180° - \angle B$.

Similarly, since quadrilateral $CDME$ is cyclic, $\angle DME = 180° - \angle DCE$. By subtraction,

$$\angle FMD = \angle FME - \angle DME = \angle DCE - \angle B. \tag{I}$$

$$\text{However, } \angle DCE = \angle BAC + \angle B. \tag{II}$$

By substituting (II) into (I), $\angle FMD = \angle BAC = 180° - \angle FAD$. Therefore, quadrilateral $ADMF$ is also cyclic and point M lies on all three circles. Point M is called the *Miquel point* of $\triangle ABC$. The points F, D, and E determine the *Miquel triangle*, $\triangle FDE$. Miquel's theorem opens the door to a variety of additional theorems,[6] one of which is presented here.

Theorem: The segments joining the Miquel point of a triangle to the vertices of the Miquel triangle form congruent angles with the respective sides of the original triangle.

Proof: Since quadrilateral $AFMD$ is cyclic (see Figures 6-34 and 6-35), $\angle AFM$ is supplementary to $\angle ADM$. But $\angle ADM$ is supplementary to $\angle CDM$. Therefore, $\angle AFM = \angle CDM$, whereupon it follows that $\angle BFM = \angle ADM$. To complete the proof, merely apply the same argument to cyclic quadrilateral $CDME$.

The Seven Circles Theorem

This remarkable result was first published in 1974.[7] This implies that there are still reasonably simple unknown results in elementary

---

[6] For more applications of Miquel's theorem, see A. S. Posamentier, *Advanced Euclidean Geometry* (Hoboken, NJ: Wiley, 2010).

[7] C. J. A. Evelyn, G. B. Money-Coutts, and J. A. Tyrrell, *The Seven Circles Theorem and Other New Theorems* (London: Stacey International, 1974), pp. 31–42.

geometry out there, waiting to be discovered by some diligent research-
ers. The configuration of the theorem is shown in Figure 6-38.

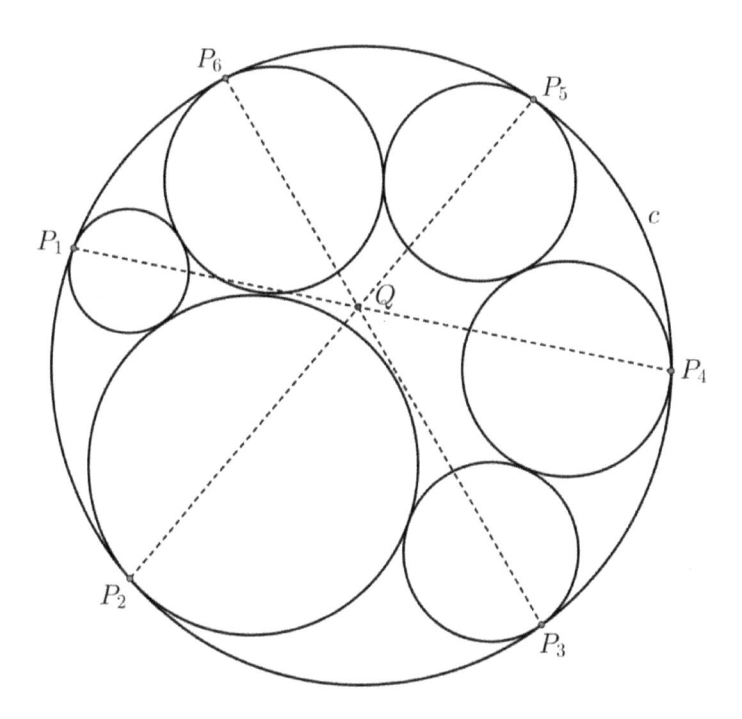

Figure 6-38

In Figure 6-38, we are given a circle c. Six more circles are also
present, each of which is tangent to c at points P_1, P_2, P_3, P_4, P_5 and
P_6, respectively, and any two successive circles among these are
also tangent to each other. In other words, the circles through P_1
and P_2 touch in a point, as do those through P_2 and P_3, those
through P_3 and P_4, those through P_4 and P_5, those through P_5 and P_6,
and also those through P_6 and P_1. If all of these pairs of circles are
tangent, it follows that the lines P_1P_4, P_2P_5, and P_3P_6 pass through a
common point Q.

This is true under quite general circumstances. In Figure 6-38, no two of these circles intersect, but as in Figure 6-39, we see that it is also true if some of these circles do intersect.

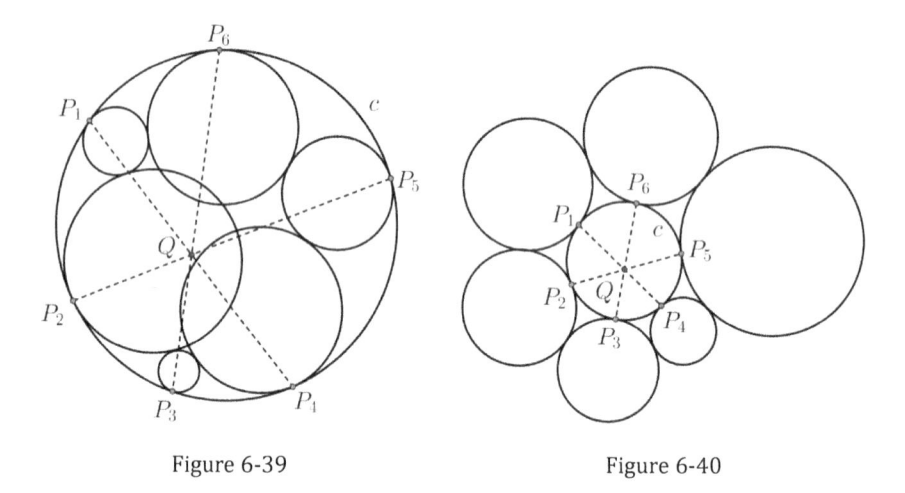

Figure 6-39   Figure 6-40

Also, as illustrated in Figure 6-40, the relationship is also true if the six tangent circles lie outside of circle c, and not inside, as in Figure 6-38. The proofs for the inside and outside cases are quite similar. Here, we will restrict ourselves to proving the inside case. In order to do this, we first have to establish a relationship concerning three pair-wise tangent circles. This is illustrated in Figure 6-41, and the statement is as follows.

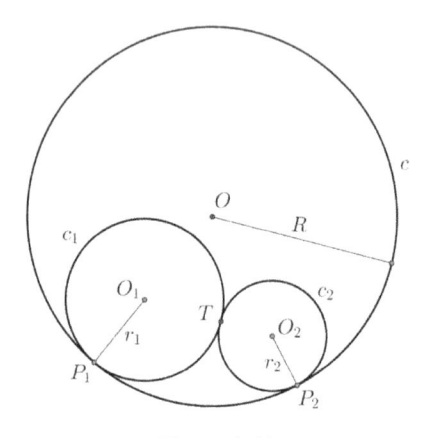

Figure 6-41

In Figure 6-41, we have c as a circle with center O and radius R. The circles c_1 and c_2 with centers O_1 and O_2 and radii r_1 and r_2, respectively, are internally tangent to circle c in points P_1 and P_2, respectively. Furthermore, circles c_1 and c_2 are externally tangent at point T. Then we have $\frac{P_1P_2^2}{4R^2} = \frac{r_1}{R-r_1} \cdot \frac{r_2}{R-r_2}$. This result can be proved in the following way. As shown in Figure 6-42, we first extend P_1T to meet circle c in A and P_2T to meet circle c in B.

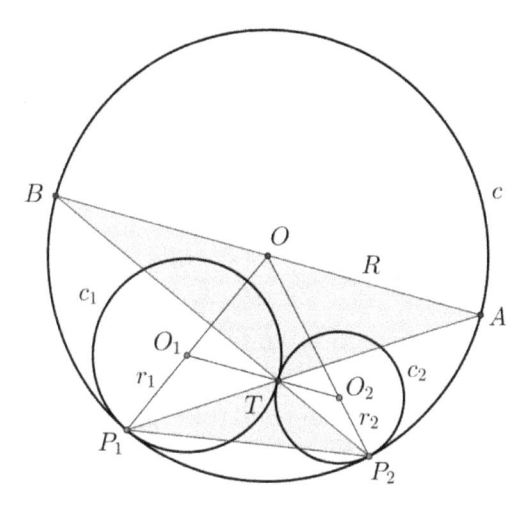

Figure 6-42

Note that $\triangle O_1P_1T$ and $\triangle OP_1A$ are both isosceles, since O_1P_1 and O_1T are both radii of circle c_1 and OP_1 and OA are both radii of circle c. These triangles have a common angle in P_1, so that we have $\angle P_1AO = \angle AP_1O = \angle P_1TO_1$. It therefore follows that OA is parallel to O_1T. Similarly, OB is parallel to O_2T since $\triangle O_2P_2T$ and $\triangle OP_2B$ are both isosceles with a common angle in P_2. Since O_1, T, and O_2 lie on a common line, so too, therefore, must A, O, and B be collinear. We can have a closer look at $\triangle TAB$ and $\triangle TP_2P_1$, which are shaded in Figure 6-42. These are similar, since $\angle ATB = \angle P_1TP_2$ are vertical angles in T, and we also have $\angle BAT = \angle BAP_1 = \angle BP_2P_1 = \angle TP_2P_1$, since $\angle BAT$ and $\angle BP_2P_1$ are both measured by the same arc BP_1. We therefore have $\frac{P_1P_2}{AB} = \frac{P_1T}{BT} = \frac{P_2T}{AT}$. Since A, O, and B are collinear, we have $AB = 2R$, and, thus,

$$\frac{P_1P_2}{2R} \cdot \frac{P_1P_2}{2R} = \frac{P_1T}{BT} \cdot \frac{P_2T}{AT} = \frac{P_1T}{AT} \cdot \frac{P_2T}{BT} = \frac{O_1P_1}{OO_1} \cdot \frac{O_2P_2}{OO_2} = \frac{r_1}{R-r_1} \cdot \frac{r_2}{R-r_2},$$

which is equivalent to our claim of

$$\frac{P_1P_2{}^2}{4R^2} = \frac{r_1}{R-r_1} \cdot \frac{r_2}{R-r_2}.$$

Using this relationship, it is now straightforward to prove the Seven Circles Theorem. Looking back to Figure 6-38, we name the circles through points P_i as circles c_i with respective radii r_i. For any two successive circles c_i and c_{i+1}, the relationship gives us $P_iP_{i+1} = 2R \cdot f(r_1) \cdot f(r_2)$, whereby the function f is defined by $f(r_1) = \sqrt{\frac{r_i}{R-r_i}}$ and $c_7 = c_1$. We therefore have that $P_1P_2 \cdot P_3P_4 \cdot P_5P_6 = 8R^3 \cdot f(r_1) \cdot f(r_2) \cdot f(r_3) \cdot f(r_4) \cdot f(r_5) \cdot f(r_6) = P_2P_3 \cdot P_4P_5 \cdot P_6P_1$ and Ceva's theorem for chords tells us that the lines, and must pass through a common point Q, as claimed.

Poncelet's Porism

Poncelet's porism[8] is named after the French mathematician Jean-Victor Poncelet (1788–1867), and the theorem further enhances our knowledge of plane geometry beyond the high school experience. The simplest version of this is illustrated in Figure 6-43.

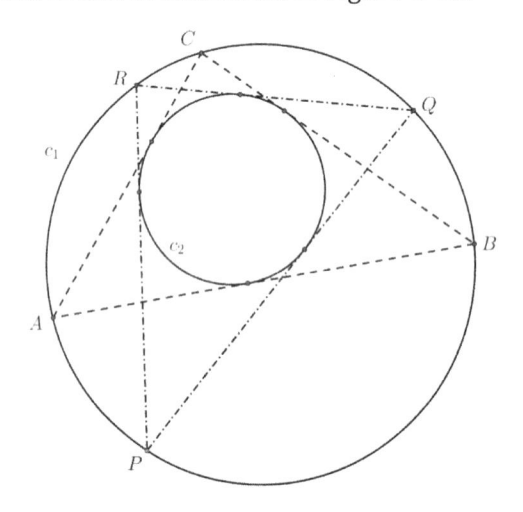

Figure 6-43

---

[8] The term *porism* is sometimes used to refer to a mathematical result that holds for some infinite class of objects, and which follows from a specific proof, in contrast to a *corollary*, which usually follows directly from a theorem.

In Figure 6-43, we see the triangle ABC and the circumcircle c_1 and incircle c_2 of this triangle. If we choose any point P on the circumcircle c_1, there exists a triangle PQR with all three vertices on circle c_1, and with all three sides tangent to circle c_2. In other words, for any such choice of a location for point P, there exists a triangle PQR sharing both its circumcircle and its incircle with the original triangle ABC.

This is actually a special case of the general result. In fact, the result is true for polygons with any number of sides, not just triangles. If a polygon with n sides and vertices has both a circumcircle and an incircle, any point can be chosen on the circumcircle, and this point will then be a vertex of a polygon with n sides and vertices that has the same circumcircle and the same incircle as the original polygon. This is illustrated for the special case $n = 4$ in Figure 6-44.

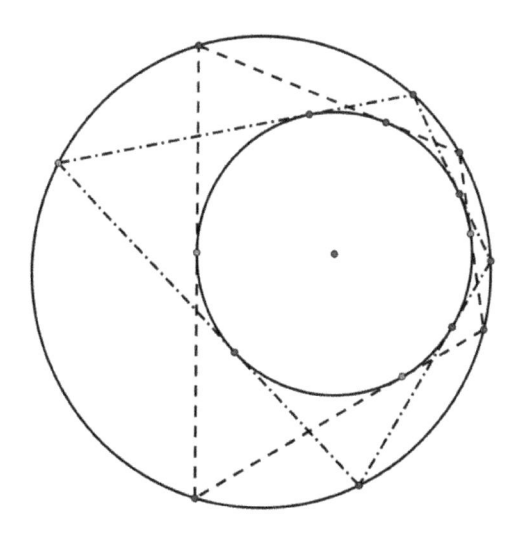

Figure 6-44

In fact, the two circles can be replaced by any conic. If a polygon with n sides and vertices has both a conic passing through all its vertices and a conic internally tangent to all its sides, any point can be chosen on the conic through the vertices, and this point will then be a vertex of a polygon with n sides and vertices that has the same circumscribed conic and the same inscribed conic as the original polygon.

Morley's Theorem

Mathematics is full of astonishing relationships. The one that we will next consider is truly amazing and perhaps just a bit beyond the typical high school experience. However, its result speaks for itself. In 1900, the English mathematician Frank Morley (1860–1937) discovered an amazing relationship that can be applied to triangles regardless of their shape. Morley demonstrated (and proved) that if we trisect each of the angles of a triangle, the intersections of the adjacent trisectors will determine an equilateral triangle. Bear in mind that what makes this so amazing is that this relationship holds true regardless of the shape of the original triangle. In Figure 6-45, we have each angle of triangle *ABC* trisected with the adjacent trisectors intersecting at points *D*, *E*, and *F*. When we join these three points, we will always arrive at an equilateral triangle *DEF*.

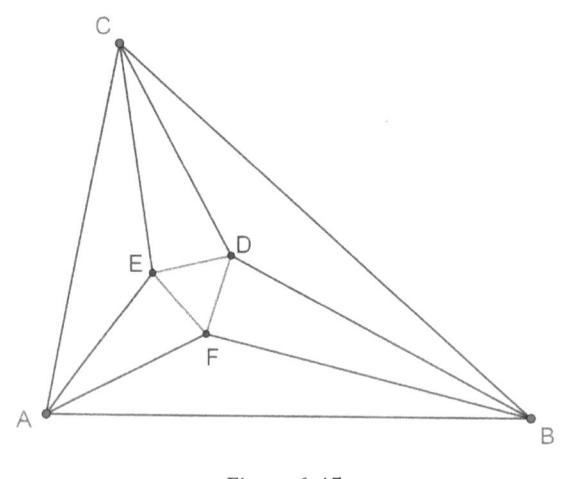

Figure 6-45

To demonstrate that Morley's theorem applies regardless of a triangle's shape, we offer a variety of triangles in Figures 6-46, 6-47, 6-48, and 6-49.

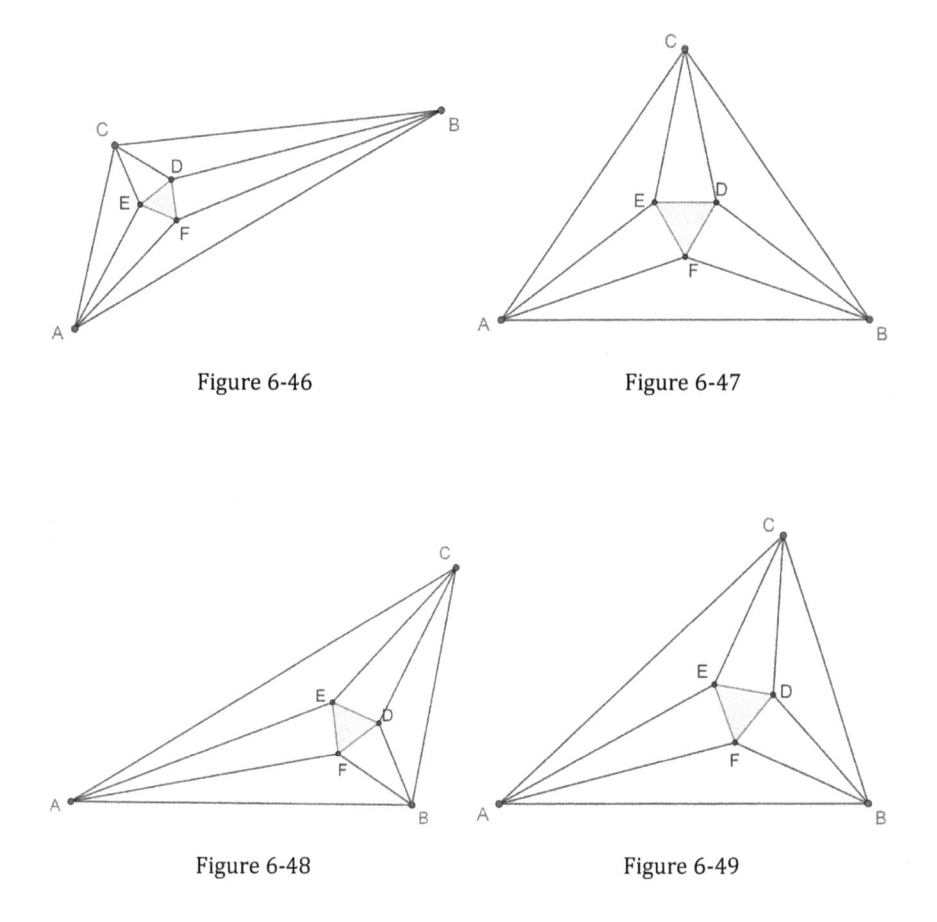

Figure 6-46   Figure 6-47

Figure 6-48   Figure 6-49

This theorem became an international challenge for mathematicians to prove. Three of these proofs are provided here.

Proof I: Let two angle trisectors at B and C (adjacent to side BC) intersect at point D and let the other two trisectors intersect at point P, as shown in Figure 6-50. Point D is clearly the incenter of $\triangle PBC$ and PD is the bisector of $\angle BPC$. From point D, two lines intersect CP and BP at points E and F, respectively, so that $\angle PDE = \angle PFD = 30°$, which is shown in Figure 6-50. We then have $\triangle PDE \cong \triangle PDF$ so that $DE = DF$. Because $\angle EDF = 60°$, we can conclude that $\triangle DEF$ is equilateral and $PD \perp EF$, as shown in Figure 6-51.

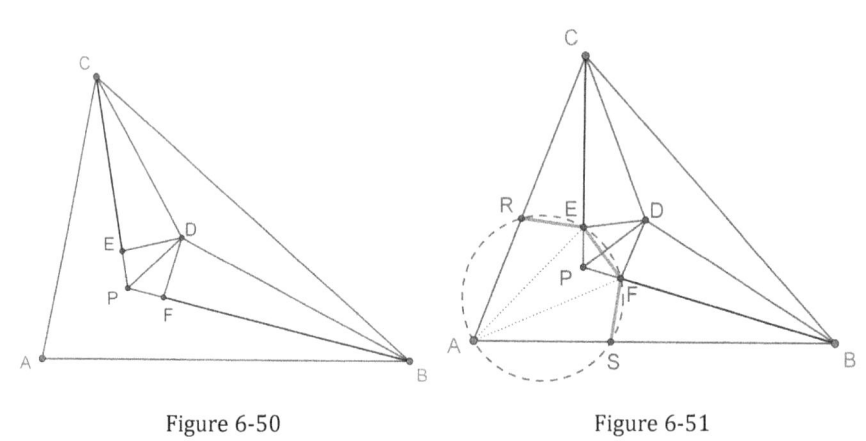

Figure 6-50   Figure 6-51

It remains to be shown that AE and AF are the trisectors of $\angle BAC$. Since CP and BP are bisectors of angle ACD and angle ABD, respectively, when we reflect D in CP and in BP, we get point R on side AC, and point S on side AB. The quadrilateral $REFS$, which can be seen in Figure 6-51 and more clearly in Figure 6-52, consists of three equal line segments RE, EF, and FS. These segments are each equal to the side length of the equilateral $\triangle DEF$, shown in Figure 6-51. This creates equal angles at E and F, which are $\angle PFS = \angle PFD = \angle PED = \angle PER$ and $\angle PFE = \angle PEF$. By addition, this results in $\angle EFS = \angle REF$. After establishing that these two angles are equal, we can also calculate their measure as follows:

$$\angle BPC = \underbrace{180°}_{\angle A + \angle B + \angle C} - \frac{2}{3}\angle B - \frac{2}{3}\angle C = \underbrace{\frac{1}{3}(\angle A + \angle B + \angle C)}_{60°} + \frac{2}{3}\angle A = 60° + \frac{2}{3}\angle A,$$

so that $\angle PFE = 90° - \frac{1}{2}\angle BPC = 60° - \frac{1}{3}\angle A$. Then, $\angle PFS = \angle PFD = 60° + \angle PFE = 120° - \frac{1}{3}\angle A$, and $\angle EFS = \angle PFE + \angle PFS = 180° - \frac{2}{3}\angle A$.

We would get the same result for $\angle REF$, but it is not necessary to do the calculation again, because we know from the above $\angle EFS = \angle REF$.

Hence, *REFS* is an isosceles trapezoid, as shown in Figure 6-51, and is *cyclic*, with the supplementary angles $\angle REF = 180° - \frac{2}{3}\angle A = \angle EFS$, and $\angle ERS = \frac{2}{3}\angle A = \angle FSA$. In trapezoid *REFS*, the diagonals are angle bisectors of $\angle R$ and $\angle S$ since $\angle RFE = \angle FRE$, resulting from isosceles $\triangle RFE$ and $\angle RFE = \angle FRS$ as $RS||EF$. Hence, we have

$$\angle RFS = \angle EFS - \underbrace{\angle RFE}_{\frac{1}{2}\angle ERS} = 180° - \frac{2}{3}\angle A - \frac{1}{2}\cdot\frac{2}{3}\angle A = 180° - \angle A.$$

This establishes that $\angle RFS$ is supplementary to $\angle A$. It therefore follows that point A lies on the circumcircle of *REFS*. Since the chords $RE = EF = SF$ are equal, they determine equal inscribed angles $\angle RAE = \angle EAF = \angle FAS$, so that AE and AF are angle trisectors of $\angle BAC$. Thus, the trisectors of the three angles of triangle *ABC* form an equilateral triangle *DEF*.

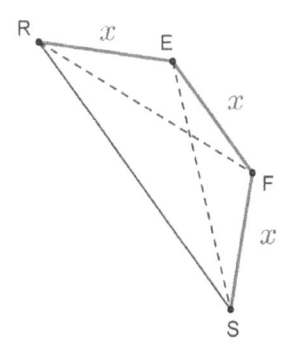

Figure 6-52

Proof II: The Japanese mathematician Yoshinori Hashimoto's original proof[9] is short and without a figure. Let α, β, and γ be arbitrary positive angles with $\alpha + \beta + \gamma = 60°$. For any angle η we put $\eta' = \eta + 60°$. Let triangle *DEF* be an equilateral triangle, and A [resp. B, C] be the point

---

[9] Y. Hashimoto, "A short proof of Morley's theorem". *Elemente der Mathematik* 62 (2007), p. 121.

lying opposite to D [resp. E, F] with respect to EF [resp. FD, DE] and satisfying $\angle AFE = \beta'$, $\angle AEF = \gamma'$ [resp. $\angle BDF = \gamma'$, $\angle BFD = \alpha'$; $\angle CED = \alpha'$, $\angle CDE = \beta'$]. Then $\angle EAF = 180° - (\beta' + \gamma') = \alpha$, and similarly $\angle FBD = \beta$, $\angle DCE = \gamma$. By symmetry it is enough to show that $\angle BAF = \alpha$ and $\angle ABF = \beta$ as well.

The perpendiculars from F to AE and BD have the same length s. If the perpendicular from F to AB has length $h < s$, then $\angle BAF < \alpha$ and $\angle ABF < \beta$. If, on the other hand, $h > s$, then $\angle BAF > \alpha$ and $\angle ABF > \beta$. Since $\angle BAF + \angle ABF = \alpha' + \beta' + 60° - 180° = \alpha + \beta$, we see that necessarily $h = s$, $\angle BAF = \alpha$, and $\angle ABF = \beta$.

Proof III: We now supplement Hashimoto's proof with Figure 6-53.

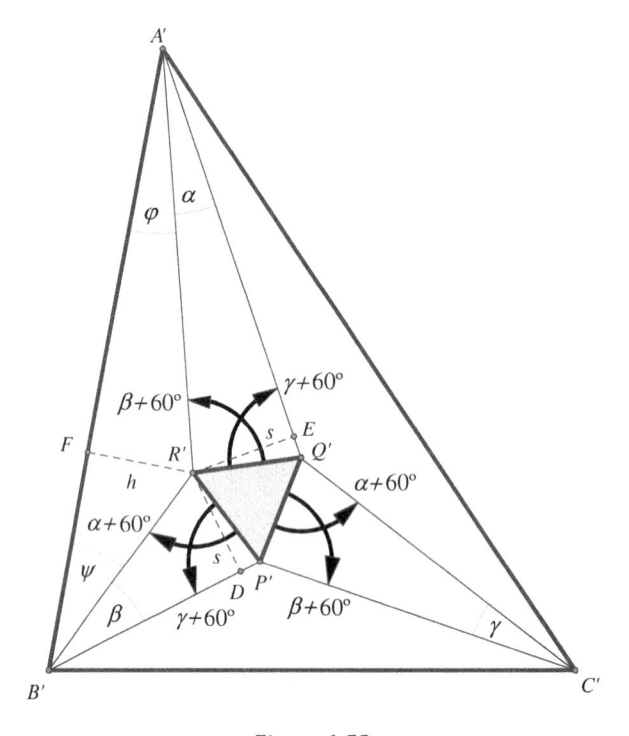

Figure 6-53

Let α, β, and γ be random angle measures, with $\alpha + \beta + \gamma = 60°$. For each of these angles we set the following: $\alpha' = \alpha + 60°$,

$\beta' = \beta + 60°$, $\gamma' = \gamma + 60°$. We have the triangle $P'Q'R'$ as equilateral. On each of the sides of the triangle $P'Q'R'$ we draw the triangles $Q'A'R'$, $R'B'P'$, and $P'C'Q'$ so that the following angles are produced:

$$\angle Q'A'R' = \alpha, \ \angle Q'R'A' = \beta' = \beta + 60°, \ \angle A'Q'R' = \gamma' = \gamma + 60°,$$

$$\angle P'B'R' = \beta, \ \angle R'P'B' = \gamma' = \gamma + 60°, \ \angle B'R'P' = \alpha' = \alpha + 60°,$$

$$\angle Q'C'P' = \gamma, \ \angle P'Q'C' = \alpha' = \alpha + 60°, \ \angle Q'P'C' = \beta' = \beta + 60°.$$

It therefore follows that $\angle Q'A'R' = 180° - (\beta' + \gamma') = 180° - (\beta + 60° + \gamma + 60°) = 60° - (\beta + \gamma) = \alpha$, and analogously, $\angle P'B'R' = \beta$, $\angle Q'C'P' = \gamma$. As a result of the symmetry, it suffices to show that $\varphi = \angle R'A'B' = \alpha$ und $\psi = \angle R'B'A' = \beta$. The perpendicular from R' to $A'Q'$ and $B'P'$ have the same length, s. If the perpendicular distance from R' to $A'B'$ is h, where $h < s$, then the following is true: $\varphi = \angle R'A'B' < \alpha$ and $\psi = \angle R'B'A' < \beta$. On the other hand, if $h > s$, then it follows that $\varphi = \angle R'A'B' > \alpha$ and $\psi = \angle R'B'A' > \beta$.

However, then $\angle A'R'B' = 360° - \angle A'R'Q' - \angle Q'R'P' - \angle P'R'B' = 360° - (\beta + 60°) - 60° - (\alpha + 60°) = 180° - (\alpha + \beta)$.

Thus, $\varphi + \psi = 180° - \angle A'R'B' = 180° - 180° - (\alpha + \beta) = \alpha + \beta$.

Therefore, it follows that $h = s$, and $\varphi = \angle R'A'B' = \alpha$, $\psi = \angle R'B'A' = \beta$.

Index

www.ingramcontent.com/pod-product-compliance
Lightning Source LLC
Chambersburg PA
CBHW050353090625
27790CB00004B/20